Simple Physics Investigations

Solutions for doing science in the classroom.

Christopher P. Garside

Seven Sides Publishing

Seven Sides Publishing of Cypress, TX, has a mission to improve teaching and the understanding of science. To contact us, send an email to simpleinvestigations@sevensidespublishing.com or visit us at sevensidespublishing.com.

Copyright © 2021 by Seven Sides Publishing and Christopher P. Garside. All rights reserved. No part of this publication may be reproduced, stored in a retrieval system, scanned, or transmitted in any form or by any means, electronic, mechanical, photocopying, recording, or otherwise, without the prior written permission of Seven Sides Publishing and Christopher P. Garside. Photocopying is permitted from this book to make copies only for the students of the teacher who owns this book; this is not for other teachers, students, or anyone else in the school or district's students.

ISBN: 9798667348016

Published by: Seven Sides Publishing, Cypress, TX.

Table of Contents

Introduction	Page 4
Unit 1 Constant Velocity	Page 17
Unit 2 Constant Acceleration	Page 28
Unit 3 Projectile Motion	Page 44
Unit 4 Force	Page 61
Unit 5 Gravitational and Centripetal Forces	Page 89
Unit 6 Momentum	Page 105
Unit 7 Energy, Work, and Power	Page 119
Unit 8 Thermodynamics	Page 149
Unit 9 Electrostatics	Page 172
Unit 10 Circuits	Page 185
Unit 11 Magnetism	Page 202
Unit 12 Waves and Sound	Page 212
Unit 13 Light	Page 234
Unit 14 Nuclear	Page 265
Physics and IPC TEKS Correlations	Page 285
Equipment List for all Investigations	Page 293

Introduction

To help teachers teach science through investigations, we have provided a series of lab manuals for Elementary Science, Middle School Science, Physics, Chemistry, Biology, Earth & Space Science, and Environmental Systems. These manuals are a rich resource for structure and investigations. There is a shortage of user-friendly labs that easily allow teachers and students to perform investigations in a timely manner. Too many labs have too much busy writing in them, where teachers do not want to take the time to read everything to figure out if it would be good for them to use with their students. If the teachers do not want to read it, do you think the students do? So I have taken a lot of the traditional labs that have been around for decades and simplified them so they are easy to read and perform. I have also added some new original labs that have never been seen before. There has been an effort to try to have teachers do more investigations with their students, but there is no plan or solution to deal with the real issues teachers have in preparing to do this. The book How to Teach Science Through Investigations has the plan, and the Simple Investigations Lab manuals have the solutions so students can learn science through investigations with minimal effort; this will make your classrooms more efficient where students learn content and practice skills at the same time. Science is a process of doing. Doing this process is the most important way for students to learn science and be able to use it in the future. We live in a culture where science-literate people are needed for jobs, but too few can be found. If you incorporate these labs with virtual labs (that I will point you to in each section of the lab manual), skill/math practice, and concept maps, you will not need to fill in gaps by giving lectures. All content can be learned through investigations and practice. Remember, we only remember 5-20% of what we hear. That 20% is when you are interested in the content. But hearing practices no science process skills and does not activate any higher cognitive thought. Lecturing is not a good option. We remember 75-80% of what we do/experience and 90-95% of what we teach. Investigations allow us to keep our students in these higher retention percentages. The main reason this works is that students spend more time in class at higher levels of Bloom's Taxonomy and stay in zones C and D of the Rigor Relevance Chart when they perform investigations. And if you add the physical way they are stimulated with the hands-on experience, you cannot deny the level of learning will be much higher while students perform investigations. This manual gives you the resources you need to teach Physics through investigations.

We separated each of these sections in the manual like you may separate your units in the class. Many studies have been done on how best to present the order of content in friendly ways. We will be doing our best to follow this type of scope and sequence. We include concept maps at the front of each section that shows the vocabulary and visual clues to how concepts relate to each other; this is a great way to organize information. It talks to the students to see how ideas work together and make it easy to chunk information to use at higher cognitive levels. At the beginning of each lab, We put the materials you will need in boldface in the beginning directions; this saves time for your lab preparation. There is also a safety question in boldface just after that for you and your students to evaluate. It says, "Looking at the material and lab we will be using, what are the safety precautions we should take to protect ourselves and materials during this investigation." Make sure to read through the lab to help you better answer this question with your students.

Virtual Labs

Hands-on labs are not the only way for students to learn science, but they are the most effective. However, many virtual labs can be used with these hands-on labs. Many investigations physically cannot be done hands-on, so some experiments will have to be done virtually. There are three sources that I have used in the past that have a good number of resources. **Physicsclassroom.com** and **PhET.colorado.edu** are free to everyone and are great to use. **Physicsclassroom.com** has teacher notes and activities/exercises that guide students through Physics Interactives. You can find them under the simulation and open, download, or print the PDF. They also have a series of Concept Builders that are a great virtual practice that can replace those worksheets that help students practice concepts, math, and skills. They can be hard to find, so above the list provided is the section where they can be found (underlined and in italics) on the website. **PhET.colorado.edu** has a variety of activities of different levels that you can explore to go with their simulations. They are easy to download and print. **ExploreLearning.com** is expensive, but its quality is much greater than the other two. When you click on a Gizmo, you can also click on lessons and find the Student Explorations that go with each Gizmo that you can modify, download, or print. They are written at a very high quality, making the students think like a scientist. At the end of each section of this lab manual, We will include a list of virtual labs from these organizations that would be great to use with these labs. Please remember virtual labs should never replace hands-on labs. If the students can learn the content live, that should be the priority because it is more of an experience that will be remembered. There are many other virtual simulations out there, but none so far have moved me to use them over the three I have mentioned here.

TIPPERs

TIPPERs are great for students to explore and think about different scenarios for each concept of Physics and Chemistry. These help students think outside the box, apply concepts to real life, and think about how multiple concepts would be applied together. I suggest you get the books of TIPPERs to practice and discuss after completing these labs and investigations.

Probe-ware

This lab manual has lots of labs that use probe-ware. Students must learn how to use probe-ware; this means teachers need to learn how to use probe-ware. Many companies use digital probe-ware with all the research, development, testing, and forensic testing they do; this has potential career opportunities that help students become more marketable for jobs if they are familiar with using probe-ware. Hooking everything up is just as easy as charging your phone. When I was a High School Science Technology Coach and researched which companies and devices would be the most user-friendly to students, I found using Vernier Probe-ware was better for high school students, but PASCO seemed better for middle school students. Both are giants in the probe-ware industry for education. Since this Lab manual and the series were written with High School in mind (many of these labs can be used for middle school classes because they are so simple) and I am more familiar with Vernier, I will be referring to Vernier Probe-ware. However, PASCO would be a great alternative.

Interfaces are devices that the probes are connected to that talk with the program (Logger Pro) that displays the data. I found the most economical and friendliest way for students to see the data from probe-ware is to use Vernier's LabQuest Mini interface hooked up to a computer with Logger Pro. LabQuest Mini has multiple ports, which is needed in many labs. They are the least expensive, so they are better on the budget. They require no batteries, so they are easy to transport if you need or want to. The other interfaces are more expensive, require batteries if you are going outside, or the stand-alone devices have a smaller screen to see the data, with less flexibility to manipulate the parameters like changing the time of data collection or changing units if you want to change or modify an experiment. There are wireless probes and interfaces (that cost more) that may be easier to use if you do not mind the cost. A computer screen is much bigger to see the data physically, so this is my preferred setup. But using any interfaces will work fine for these labs.

Connecting the Probe-ware

To hook them up, you will plug your probe into one of the channels or the sonic on the interface. If the plug does not fit in smoothly, either you are plugging it in upside-down or in the wrong port. Then take the little chord that looks like it would go into your phone and plug that into your interface. Take the other end, and plug it into a USB port in your computer. Open up Logger Pro on your computer. If everything is hooked up properly and the computer and interface are working properly, you will see a green button at the top of the computer screen that says collect. Many of the labs have preset settings in Logger Pro. You will use the manila folder at the top left of the toolbar in Logger Pro to find the folders and files you will be instructed to go to for these specific settings for different labs. Whenever you get the physical equipment, they will have detailed instructions in the box they come in on how to hook them up if you are still confused. They will also have instructions on how to calibrate the probes if needed. A few probes require frequent calibration; if we use any, it will be discussed in the lab directions. The more you use probe-ware, the easier it gets to set up. I usually only have to show my students twice for them to be able to set the equipment up on their own. But as you are showing them, have them physically do it. You can also find detailed instructions online at Vernier.com. Many more detailed labs can also be found there under lab ideas.

You also can use standard equipment like spring scales for force sensors or thermometers for temperature probes. Because schools want to integrate more technology, We wrote these labs to use probe-ware wherever applicable. Because they are so simple, these labs can be modified to fit whatever equipment you have. There are very few labs that I have used in my career that I did not modify how I presented them. One reason we wrote these labs this way was to customize them to the Texas TEKS and National Standards. We also wrote them the way we thought teachers would want to use them.

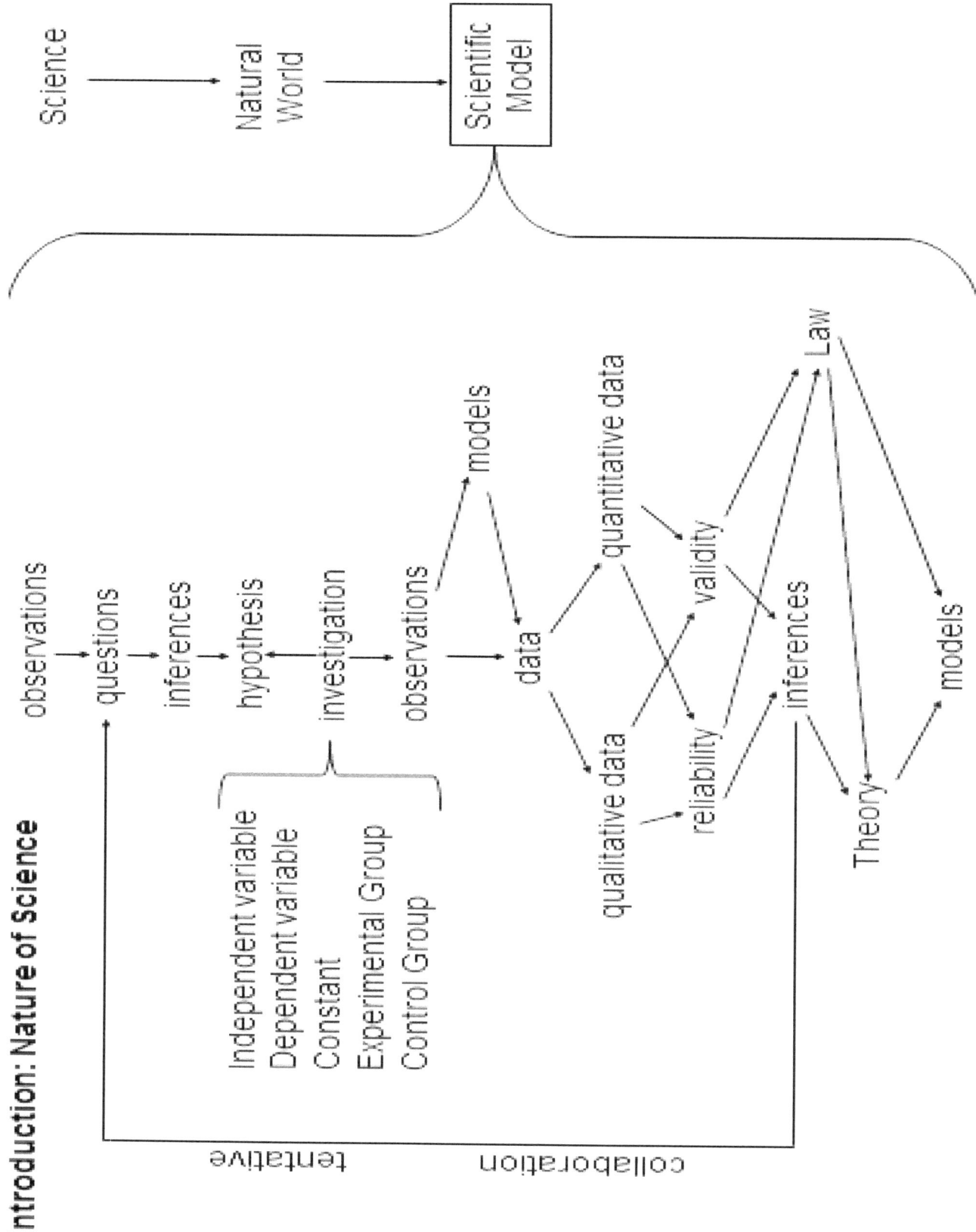

Focus on the Process

Directions:

Get a **small Legos set**. Teachers, make sure it is not too easy for your students. You are going to try to put it together in two different ways. Time how long it takes to put it together each way and answer the questions that follow. **Looking at the materials and lab we will be using, what are the safety precautions we should take to protect ourselves and materials during the investigation?**

    A) Take the Lego pieces and try to construct the picture (the **product**) on the box's front cover. Look at nothing but the front cover and the Lego pieces.

    B) When 20 minutes have passed, or you are done, take what you have made totally apart. Take out the directions (the **process**) and construct the product while using the step-by-step directions. Time how long it takes to complete the set.

Questions:

1) How did it feel to try to construct the Legos (A) without any directions?

2) Did you finish? If so, how long did it take?

3) How did it feel to construct the Legos (B) with the step-by-step directions?

4) Did you finish? If so, how long did it take?

5) Which strategy (A or B) allowed you to complete the product?

6) Which strategy (A or B) was more intimidating?

7) Which strategy (A or B) allowed you to see what is under the surface?

8) Which strategy (A or B) will allow you to learn more?

We often get anxious or procrastinate when faced with a large task. We are tempted to take a "shortcut" (copy or cheat, we do not learn much when we do this). There are pain and stress hormones that are released when this happens. One way to overcome this is to just worry about the next step in the process and not worry about the product. You can see and measure progress, which makes the process not feel too bad. Another way is to just start working. When you start working, those pain and stress hormones stop getting released so that anxiety goes away; this is why when we want to learn efficiently and effectively, we must:

Focus on the _____ and the _____ will take care of itself.

9) How is putting the Lego pieces together like putting ideas together to understand concepts?

Measurement Lab

Directions:

You will need **water**, a **scale**, a **meter stick**, a **temperature probe** attached to an **interface** connected to a **computer** with **Logger Pro**, a **100 mL graduated cylinder**, and a **stopwatch**. **Looking at the materials and lab we will be using, what are the safety precautions we should take to protect ourselves and materials during the investigation?**

1) Take the graduated cylinder and find its mass empty; write this in Data Table 1.
2) Add 50 mL of water to the graduated cylinder. Make sure you use the meniscus properly where the volume is at the bottom of the meniscus. Have the teacher check that you measured it properly. Have each person in your group empty and fill the graduated cylinder with 50 mL of water. As they do so, have each person time how long it takes to fill the graduated cylinder and check it is correct (it is not a race, just a chance to get familiar with using the graduated cylinder and stopwatch).
3) Now find the mass of the graduated cylinder with the 50 mL of water in it. Subtract the empty graduated cylinder's mass from the full and write the water's mass in Data Table 1 below.
4) Hook your temperature probe up to an interface and hook your interface up to a computer with Logger Pro (unless you have a LabQuest 2, then just hook your probe to the LabQuest 2). Find where the Logger Pro is located on your computer so you can use it again in the future. Once open, find the graduated cylinder's water temperature in both Fahrenheit and Celsius (you will have to figure out how to change units). Write these data in Data Table 1.
5) Take your meter stick and measure the length of the graduated cylinder. Measure the width of the base in centimeters. Write these data in Data Table 1.

Data Table 1

Object	Mass (g)	Volume (mL)	Time to Fill (s)	Temp (°F)	Temp (°C)	Length (cm)	Width (cm)
Graduated Cylinder		✗		✗	✗		
Water			✗			✗	✗

Questions:

1) Convert a length to meters, the volume to liters and a mass to kilograms, and Celsius to Kelvin.

   Length _____ m   Volume _____ L   Mass _____ kg   Temp _____ K

2) What do you notice about the mass of the water compared to its volume?

3) What can happen to your investigations if your measurements are not accurate or precise?

4) Why do you think the rest of the world uses the metric system over the English system?

Simple Physics Investigations

Seven Sides Publishing

Patterns in Pennies

Directions:

You will need a **ruler**, 10 **pennies**, a **balance**, a **roll of pennies**, and an **empty penny roll**. Looking at the materials and lab we will be using, what precautions should we take to protect ourselves and materials during the investigation?

1) Find the mass of one penny with a scale to the nearest .1 g. Then measure the height of the penny in millimeters. Write these in Data Table 1 below.
2) Place another penny on top of the original penny and find the mass and height of the two pennies. Write these in Data Table 1 below.
3) Keep adding pennies one by one, measuring the mass and height until you have 10 pennies on the scale.
4) Make a line graph with the mass on the (*x*) axis and the height on the (*y*) axis for the pennies on Graph 1.

Data Table 1

Number of Pennies	Mass	Height
1		
2		
3		
4		
5		
6		
7		
8		
9		
10		

Graph 1

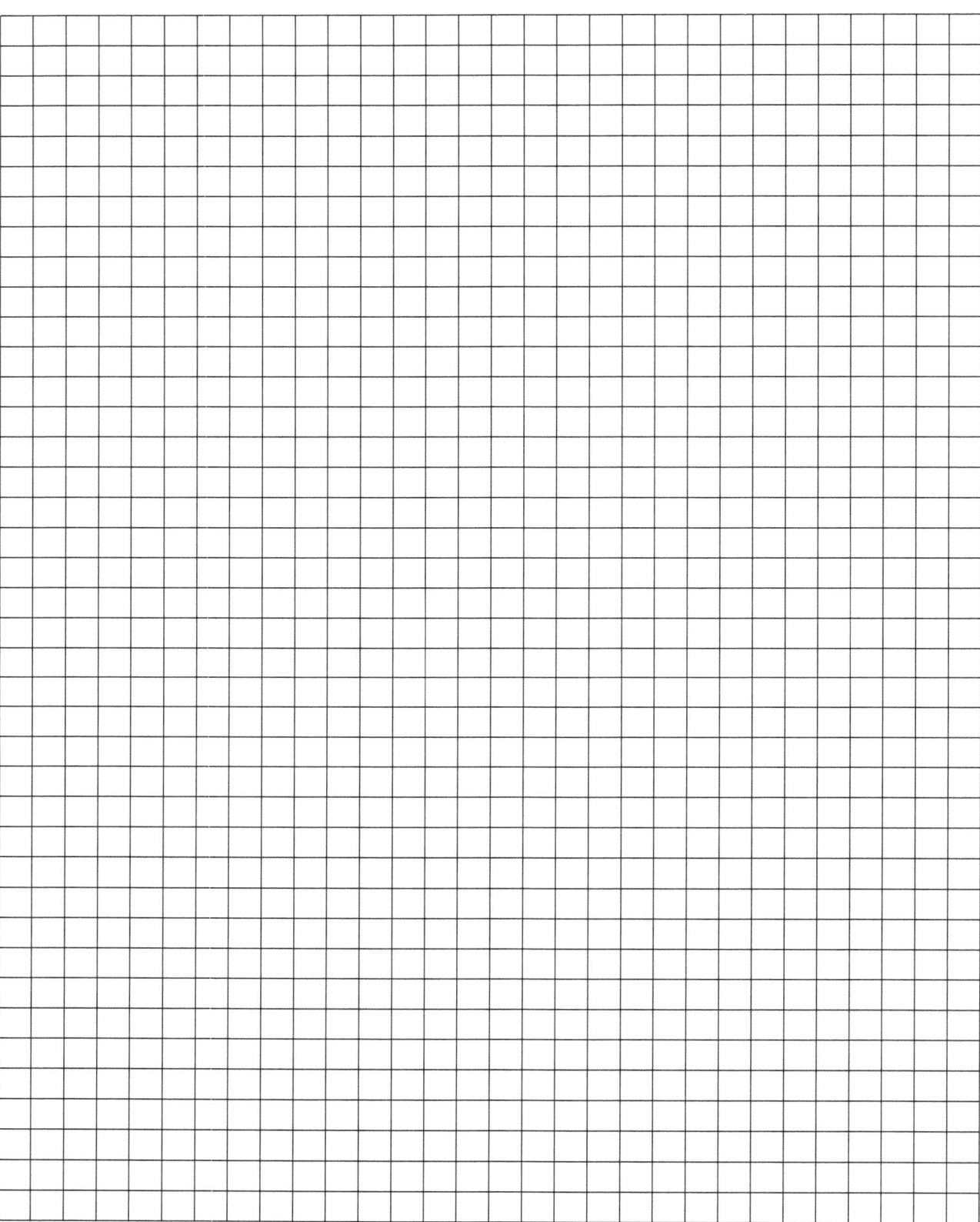

Questions:

1) What do you notice about the graph?

2) Is this a direct or inverse relationship between mass and height?

3) Do all pennies have the same mass? (Explain)

4) Do all the pennies have the same thickness? (Explain)

5) Use your data to estimate how many pennies are in the coin roll. How many pennies do you think are in the roll?

6) What did you do to estimate the number of coins?

7) What else could you do to estimate the coins?

8) Try your answer to #7. Do you get the same number as #5?

9) Carefully open up the coin roll and find out how many pennies there are. How close were you to the real number? After being done counting, carefully close the roll back up.

10) Calculate the % accuracy by taking the lowest number between your guess and the actual number dividing by the higher of the two, and then multiplying by 100.

11) What were some sources of error?

Virtual Investigations that go with Introduction

ExploreLearning.com:

    Unit Conversions Gizmo

    Graphing Skills Gizmo

    Elevator Operator Gizmo

    Measuring Volume Gizmo

    Weight and Mass Gizmo

    Triple Beam Balance Gizmo

    Reaction Time 1 Gizmo

    Reaction Time 2 Gizmo

Physicsclassroom.com:

    Concept Builders:

        Relationships and graphs

        Experiments and Variables

        Proportional Reasoning

        Using Graphs

        Which One Does Not Belong

        Chemistry

        Significant Digits and Measurement

        Metric System

        Metric Estimation

Unit 1 Constant Velocity — Motion

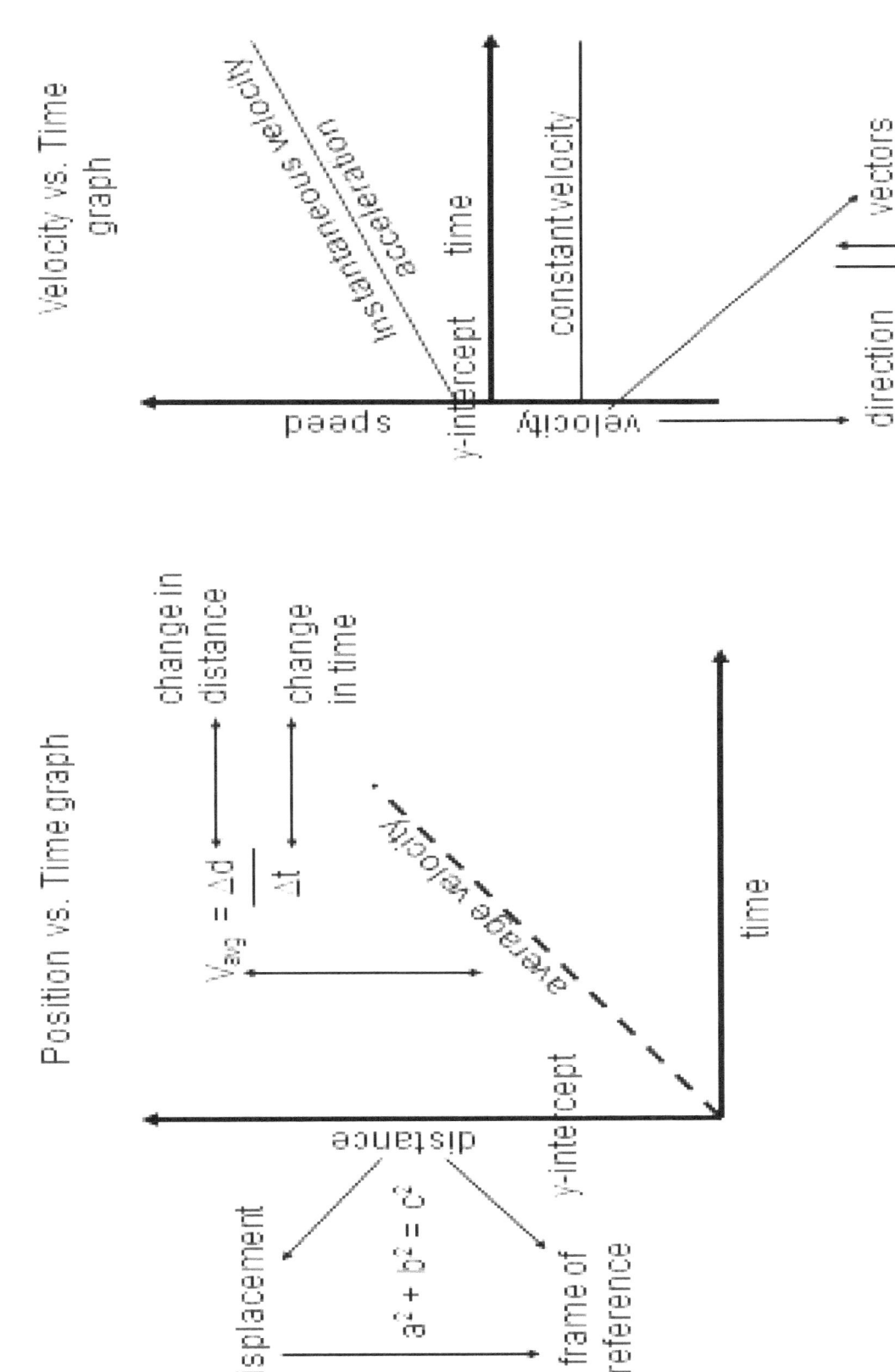

The Motion of a Toy Car

Directions:

You will need an **electric toy car** (that moves at a constant speed), five **stopwatches**, and **masking tape** to mark each meter with a **meter stick** on the floor for 5 meters. **Looking at the materials and lab we will be using, what are the safety precautions we should take to protect ourselves and materials during the investigation?**

1) Have five students line up one meter apart, each with a stopwatch.
2) Everyone starts the stopwatch when the toy car moves past the starting line.
3) When the car passes each person at their meter line, they need to stop their stopwatch.
4) Record the times in the data table below.
5) Clear the stopwatches and repeat the procedure in # 2-4. Then calculate the averages for each distance by adding the two times and dividing by two.
6) Graph the averaged data in the Time vs. Distance graph on the next page.
7) Find the graph's slope by taking the rise (distance) divided by the run (time).

Data Table 1

Distance (m)	Trial 1 Time (s)	Trial 2 Time (s)	Average Time (s)
0 m	0 s	0 s	0 s
1 m			
2 m			
3 m			
4 m			
5 m			

Graph 1

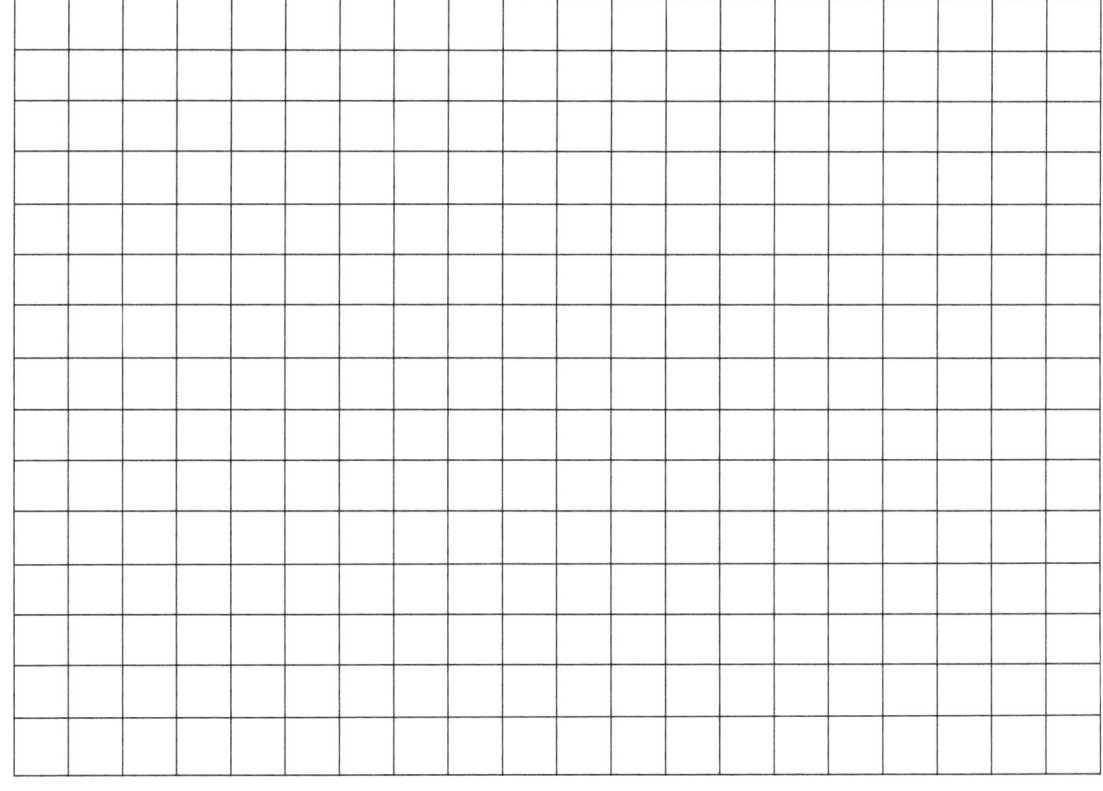

Questions:

1) On the distance-time graph, what does the slope of the line graph tell you?

2) What do you think a flat/horizontal line will tell you on a distance-time graph?

3) What was the average speed of the toy car?

4) Was the speed of the car constant or changing?

5) How far did the car travel while it was being timed?

6) What was its displacement?

7) How are distance and displacement related?

Simple Physics Investigations — Seven Sides Publishing

The Motion of a Bowling Ball

Directions:

You will need a **bowling ball** (an **air puck** can also be used), five **stopwatches**, a **large pillow** or something to act as a bumper, and **masking tape** to mark each meter with a **meter stick** on the floor for 5 meters. **Looking at the materials and lab we will be using, what are the safety precautions we should take to protect ourselves and materials during the investigation?**

1) Have five students line up one meter apart, each with a stopwatch.
2) Set up the bumper to stop the bowling ball at the far end.
3) Everyone starts the stopwatch when the bowling ball moves past the starting line.
4) When the ball passes each person at their meter line, they need to stop their stopwatch.
5) Record the times in Data Table 1 below.
6) Clear the stopwatches and repeat the procedure in # 3-5. Then calculate the averages for each distance by adding the two times and dividing by two.
7) Graph the averaged data in the Time vs. Distance graph on the next page.
8) Find the graph's slope by taking the rise (distance) divided by the run (time).

Data Table 1

Distance (m)	Trial 1 Time (s)	Trial 2 Time (s)	Average Time (s)
0 m	0 s	0 s	0 s
1 m			
2 m			
3 m			
4 m			
5 m			

Graph 1

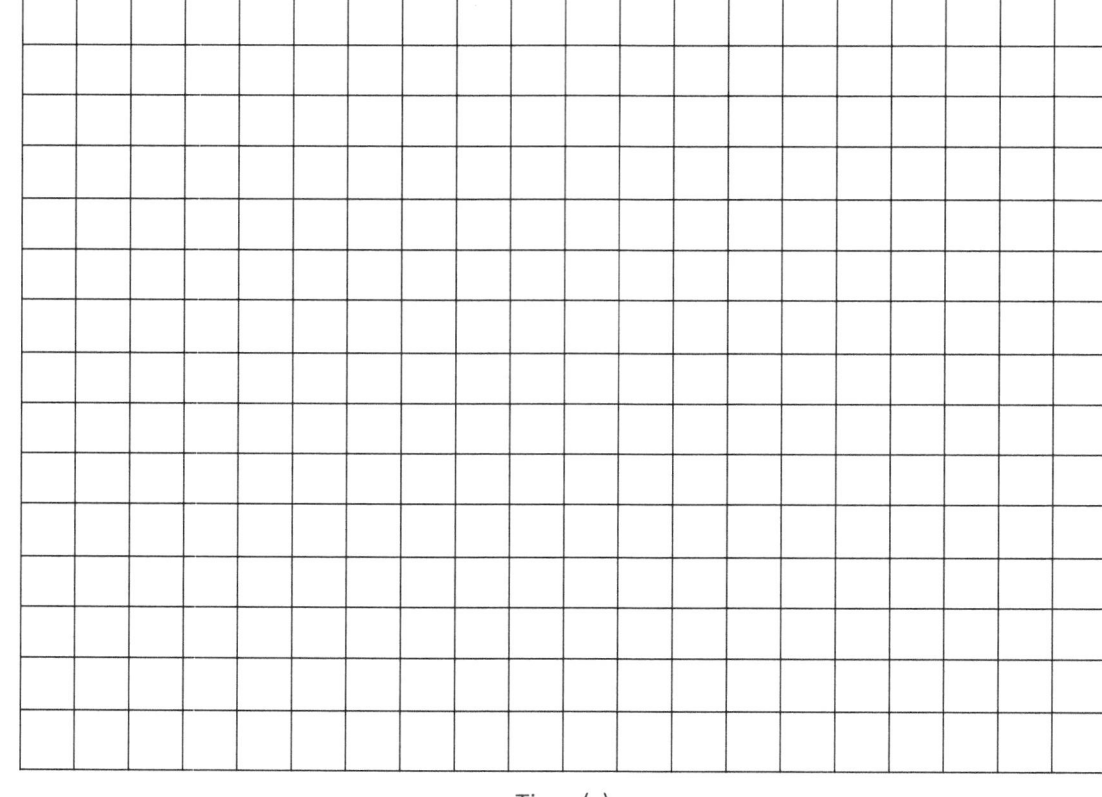

Questions:

1) On the distance-time graph, what does the slope of the line graph tell you?

2) What do you think a flat/horizontal line will tell you on a distance-time graph?

3) Imagine a bowling ball dropped from a great height. How would the motion of this ball relate to the one in the lab?

4) What was the average speed of the bowling ball?

5) Did the speed of the ball seem constant or changing?

6) How far did the bowling ball travel while it was being timed?

7) What was its displacement?

8) How are distance and displacement related?

Motion Detector Lab

Directions:

Below are some distance-time graphs. You need a **flat board** in front of a **motion detector** attached to an **interface** connected to a **computer** with **Logger Pro**. Press "Collect" to start collecting data with the motion detector to try and simulate the graphs seen below by using the flat board by moving it towards and away from the motion detector. When you have successfully made the first graph, fill in the row of data for that graph describing the motion. Then move on to the next graph until you have successfully made each graph below. **Looking at the materials and lab we will be using, what are the safety precautions we should take to protect ourselves and materials during the investigation?**

Graphs to Make	Graph section	Motion (stationary, toward, or away)	Velocity (+,-, or 0)	Which section moved the fastest?
1.	A	S T A	+ - 0	A B
	B	S T A	+ - 0	
2.	A	S T A	+ - 0	
	B	S T A	+ - 0	A B C
	C	S T A	+ - 0	
3.	A	S T A	+ - 0	
	B	S T A	+ - 0	A B C
	C	S T A	+ - 0	
4.	A	S T A	+ - 0	A B
	B	S T A	+ - 0	

On the next page, make a Motion Map and Velocity vs. Time Graph for each of the graphs you created above.

	Motion Map	Velocity vs. Time Graph
1		
2		
3		
4		

Questions:

1) On a distance-time graph, what does a horizontal line represent?

2) On a distance-time graph, what does a positive slope represent?

3) On a distance-time graph, what does a negative slope represent?

4) What does a steeper slope represent?

5) Explain how to draw a motion map.

6) Explain how to draw a velocity-time graph.

Virtual Investigations that go with Constant Velocity

ExploreLearning.com:

    Measuring Motion Gizmo

    Distance-Time Graphs Gizmo

    Distance-Time and Velocity-Time Graphs Gizmo

    Vectors Gizmo

    Adding Vectors Gizmo

    Pythagorean Theorem Gizmo

    Pythagorean Theorem with Geoboard Gizmo

    Cat and Mouse (Modeling with Linear Systems) Gizmo

    Distance Formula Gizmo

PhET.colorado.edu:

    Vector Addition

    Maze Game

    Motion 2D

    The Moving Man

    Ladybug Motion

Physicsclassroom.com:

    Physics Interactives:

        Vector Walk

        Vector Addition

        Name that Vector

        Vector Guessing Game

        Vector Addition: Does Order Matter?

Riverboat Simulation

Concept Builders:

Kinematics

Distance vs. Displacement

Speed-Distance-Time

Motion Diagrams

Position – Time Graphs – Conceptual Analysis

Position – Time Graphs – Numerical Analysis

Vectors and Projectiles

Vector Direction

Head – To – Tail Vector Addition

Vector Addition

Component Addition

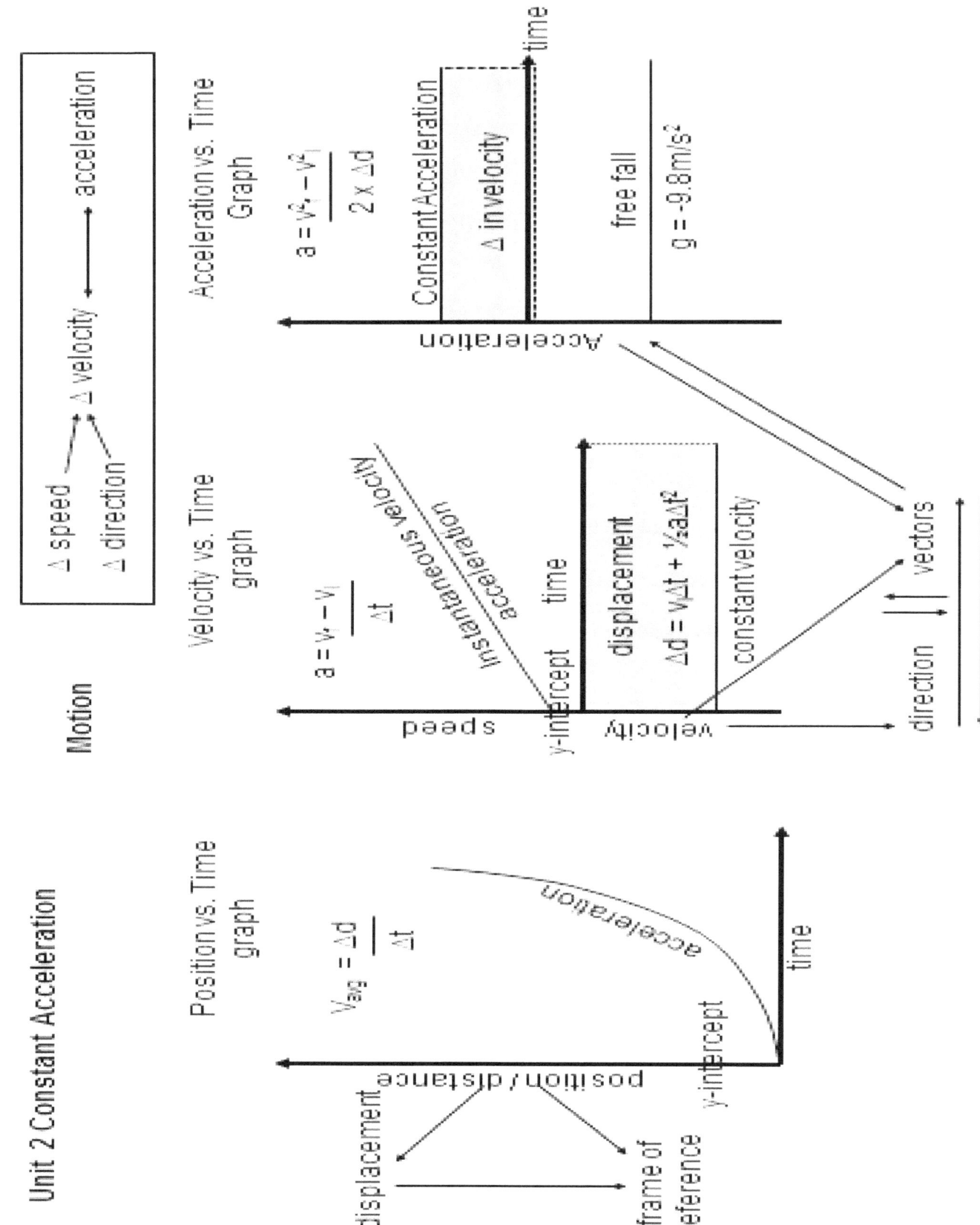

Simple Physics Investigations Seven Sides Publishing

Marbles in Motion

Directions:

Get a segment of **hot wheels track**, **small stickers**, a **stopwatch**, and a **marble**. Have something that one side of the track can be placed on to raise that end to create a ramp. **Looking at the materials and lab we will be using, what are the safety precautions we should take to protect ourselves and materials during the investigation?**

1) Set the ramp so the ramp's bottom is along the edge of a tile on the floor. Most tiles in schools are 1 foot in length. Clear a path for 5 feet.
2) Adjust the height of the ramp so that the marble will just make it past 5 feet.
3) Place small stickers on the floor at the ramp's base and each foot past the base. The last one is 5 feet away from the base of the ramp.
4) Place a small sticker on the ramp to mark where you will place your marble for each trial to let it roll down the ramp; this keeps the distance your marble will be accelerating down the ramp constant.
5) Place your marble on the ramp and let it roll down (do not push). Time with a **stopwatch** how long it takes for the marble to move from the ramp's base to one foot away.
6) Repeat #5 four more times. Record the data in Data Table 1 below.
7) Repeat #s 5 and 6 for the distances 2 feet, 3 feet, 4 feet, and 5 feet away.
8) Find the average time for each distance.
9) Then calculate the average velocity for each distance by taking the distance and dividing it by the average time and write that in Data Table 1.

Data Table 1

Trial	1 foot	2 feet	3 feet	4 feet	5 feet
1					
2					
3					
4					
5					
Average Time					
Average Speed					

10) Take the average speed and plot them on the graph to make a speed-distance graph; this will look similar to a velocity-time graph since the longer the distance, the more time it takes. The shape we see will be the same as looking at accelerated motion on a velocity-time graph.
    a. This graph will be the same shape we would see for constant velocity motion on a distance-time graph.

Velocity vs. Distance Graph

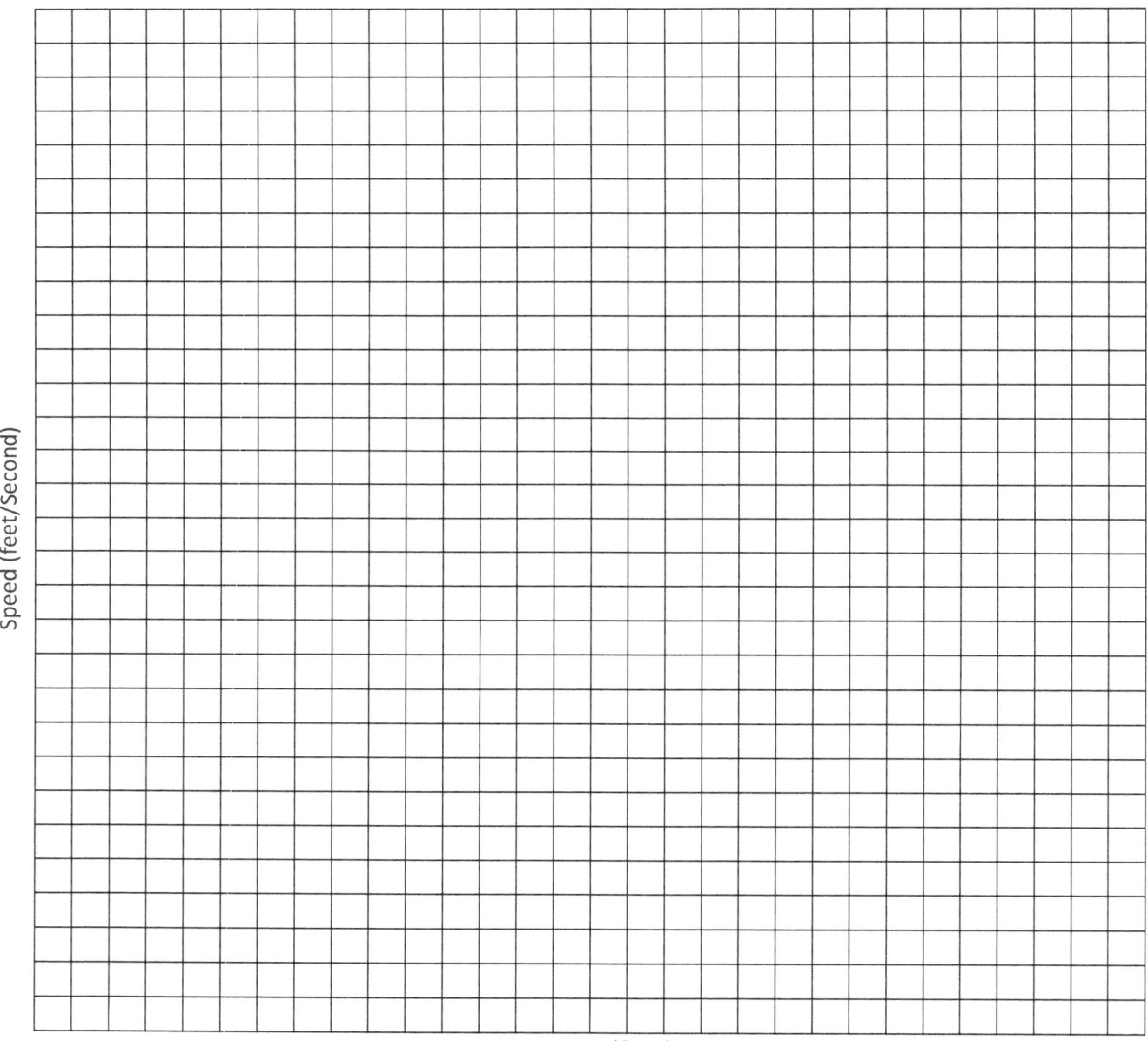

Questions:

1) Describe the motion of the marble as it moved down the ramp. If you have to, place the marble on the ramp and let it go to observe it move down the ramp.

2) What force caused the marble to speed up on the ramp?

3) How could we make the marble have a higher velocity at the bottom of the ramp?

4) Describe the motion of the marble as it moved across the floor. Place the marble on the ramp again and observe it roll across the floor if you have to.

5) What caused the marble to slow down across the floor?

6) How could we make the marble slow down faster across the floor?

7) At what distance would the average velocity be happening? At this distance would be the instantaneous velocity that is the same as the average velocity.

8) What was the shape of this graph? This shape of the graph is what acceleration looks like on a velocity-time graph.

9) What conditions would you need for the marble to have no positive or negative acceleration?

Observing Changes in Motion

Directions and Questions:

You will need a **ball** to throw up and down and bounce off objects. You will need a safe place to allow the ball to bounce in different directions. **Looking at the materials and lab we will be using, what are the safety precautions we should take to protect ourselves and materials during the investigation?**

1) **Acceleration** is defined as a change in motion. That can be a change in **velocity** or **direction**; it can also be a change in **speed**. How are velocity and speed different?

    a. How are they the same?

2) Gently throw your ball up in the air. What do you observe that shows it is accelerating?

3) Is it positive or negative acceleration?

4) Is acceleration changing while it is in the air? Explain why.

5) What is the acceleration of your ball?

6) Gently throw your ball and bounce it off a wall. How did the ball change motion?

7) Was this object accelerating? Give evidence for your answer.

8) Now gently throw your ball and let it bounce around. What eventually happened to your ball?

9) What do you think caused this to happen to your ball?

10) Was acceleration constant or changing? Explain your answer.

   a. Was it positive or negative acceleration?

11) When was your ball accelerating positively today?

12) When did your ball have constant acceleration?

13) What things determine whether something is accelerating positively or negatively?

14) How did you see the ball change its motion today?

Motion in Real Life

Directions and Questions:

Go outside and observe four objects in motion (two with constant velocity and two with acceleration) and answer the following questions.

1) What did you see that had constant velocity?

   a. Describe its motion.

2) What did you see that had acceleration?

   a. How did it accelerate? Describe its motion.

      i. Was it positive or negative acceleration? Explain.

   b. What forces do you think caused it to accelerate?

3) What else did you see that had constant velocity?

   a. Describe its motion.

4) What else did you see that had acceleration?

   a. How did it accelerate? Describe its motion.

      i. Was it positive or negative acceleration? Explain.

   b. What forces do you think caused it to accelerate?

Ball Bounce

Directions:

You will need to get a **small wire basket**, a **large bouncy ball**, and attach a **motion detector** to an **interface** connected to a **computer** with **Logger Pro**. **Looking at the materials and lab we will be using, what are the safety precautions we should take to protect ourselves and materials during the investigation?**

1) Take the small wire basket and place it over the motion detector, so the ball (large bouncy ball) that you drop will not hit it. The spacing of the wires on the basket needs to be wide enough not to be detected by the motion detector.

2) When placing the basket over the motion detector, make sure the basket's gap is directly over the sensor, so the motion detector will only see the ball.

3) In Logger Pro, open folder Physics with Vernier and file #06 Ball Toss.

4) Press "Collect" and drop the ball, letting it bounce on the basket. It is best if the ball bounces a few times to see what is happening in the data.

5) Look at the graphs and have the students see where each bounce is. Notice for each bounce; the graph sizes get less and less. Why would that happen?

6) Move the display to see only one bounce for all three graphs. Use that display to label the ball's motion on all three graphs simultaneously. Students can use the picture on the next page (similar to what you should see) to label what happened in the graphs they made.

7) Place a label on the graph where the ball is at the highest point, where the ball moves up, where the ball moves down, and where the ball is in contact with the basket.

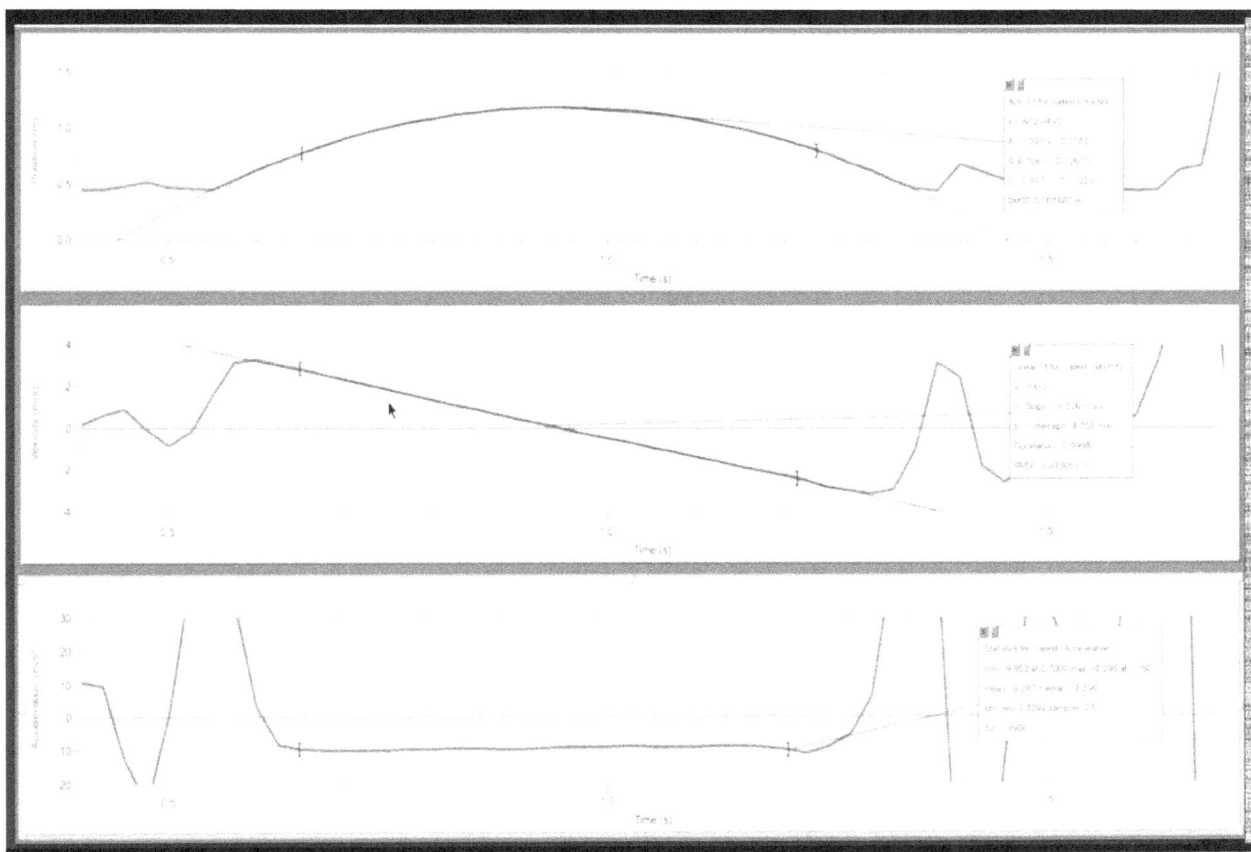

Questions:

1) What type of motion is this?

2) What is the shape of the graph of the ball as it goes through negative acceleration for:
    a. Distance vs. Time graph

    b. Velocity vs. Time graph

    c. Acceleration vs. Time graph

3) Compare this with the shape of constant velocity for:
    a. Distance vs. Time graph

b. Velocity vs. Time graph

c. Acceleration vs. Time graph

4) Make sure you study this lab to tell the difference between constant velocity and acceleration graphs.

Simple Physics Investigations   Seven Sides Publishing

Cart on a Ramp

Directions:

You will need to get a **spring cart** with **Vernier's Dynamics System** with a wall, and a **motion detector** attached to an **interface** connected to a **computer** with **Logger Pro**. **Looking at the materials and lab we will be using, what are the safety precautions we should take to protect ourselves and materials during the investigation?**

1) Set up the spring cart on a ramp like one used in Vernier's Dynamics System, where one end of the ramp is propped up in the air.

2) Place the motion detector on the top end of the ramp with the sensor facing down the ramp.

3) In Logger Pro, open the folder Physics with Vernier and file #03 Cart on a Ramp.

4) Hold the cart at the top of the ramp with the spring side down and press "Collect," and let go of the cart, allowing it to roll down the ramp and bounce off the wall. It is best if the cart bounces a few times to see what is happening in the data.

5) Look at the graphs and have the students see where each bounce is. Notice for each bounce; the graph sizes get less and less. Why would that happen?

6) Move the display to see only one bounce for all three graphs. Use that display to label the cart's motion on all three graphs simultaneously. Students can use the picture on the next page (similar to what you should see) to label what happened in the graphs they made.

7) Place a label on the graph where the cart is at the highest point, where the cart moves up, where the cart moves down, and where the cart is in contact with the ramp's bottom wall.

Lab3Part1

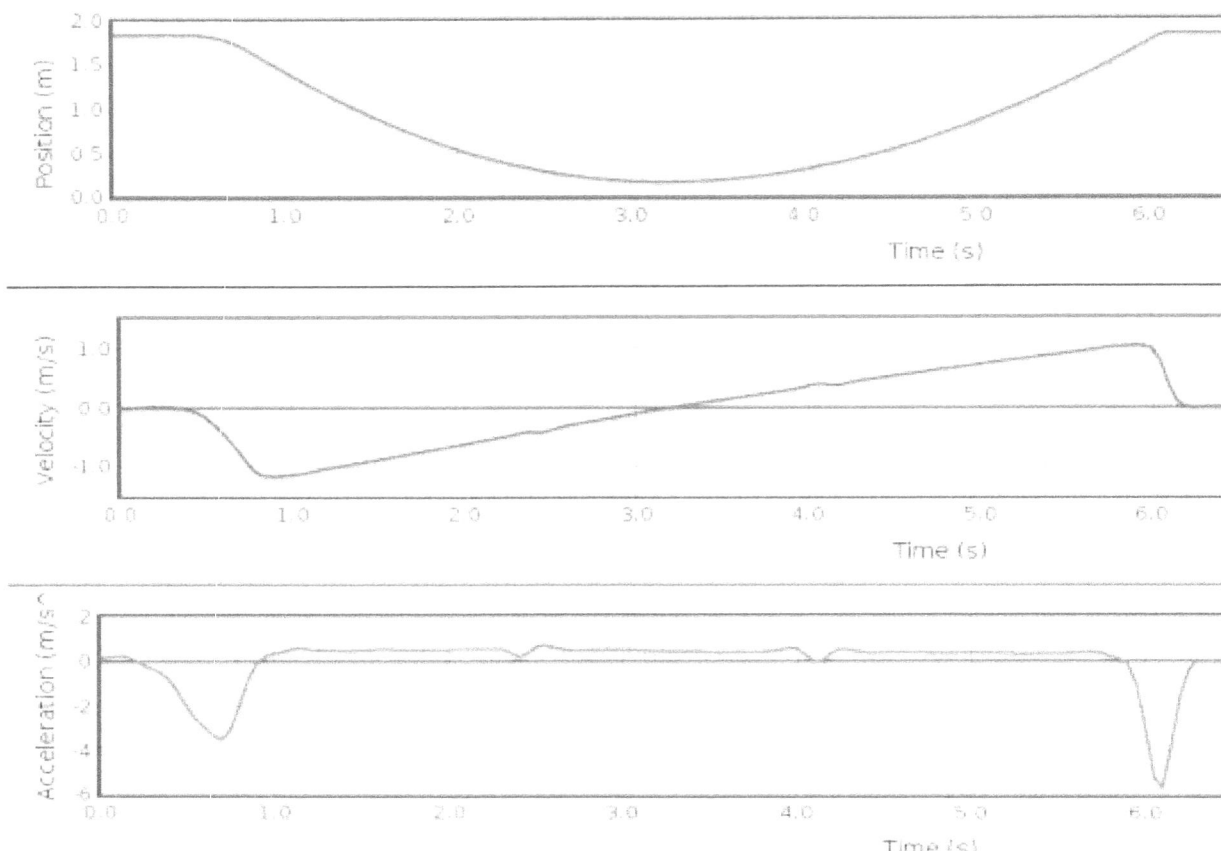

Questions:

1) What type of motion is this?

2) What is the shape of the graph of the cart as it goes through positive acceleration for:
   a. Position vs. Time graph

   b. Velocity vs. Time graph

   c. Acceleration vs. Time graph

3) Compare this with the shape of constant velocity for:
    d. Distance vs. Time graph

    e. Velocity vs. Time graph

    f. Acceleration vs. Time graph

4) Make sure you study this lab to tell the difference between constant velocity and acceleration graphs.

Virtual Investigations that go with Constant Acceleration

ExploreLearning.com:

    Free Fall Tower Gizmo

    Sled Wars Gizmo

    Atwood Machine Gizmo

PhET.colorado.edu:

    Maze Game

    Motion 2D

    The Moving Man

    Ladybug Motion

Physicsclassroom.com:

    Physics Interactives:

        Kinematics

        Graphs & Ramps

    Concept Builders:

        Kinematics

        Acceleration

        Name That Motion

        Motion Diagrams

        Graph That Motion

        Match That Graph

        Velocity – Time Graphs

        Dots and Graphs

Words and Graphs

Which One Doesn't Belong

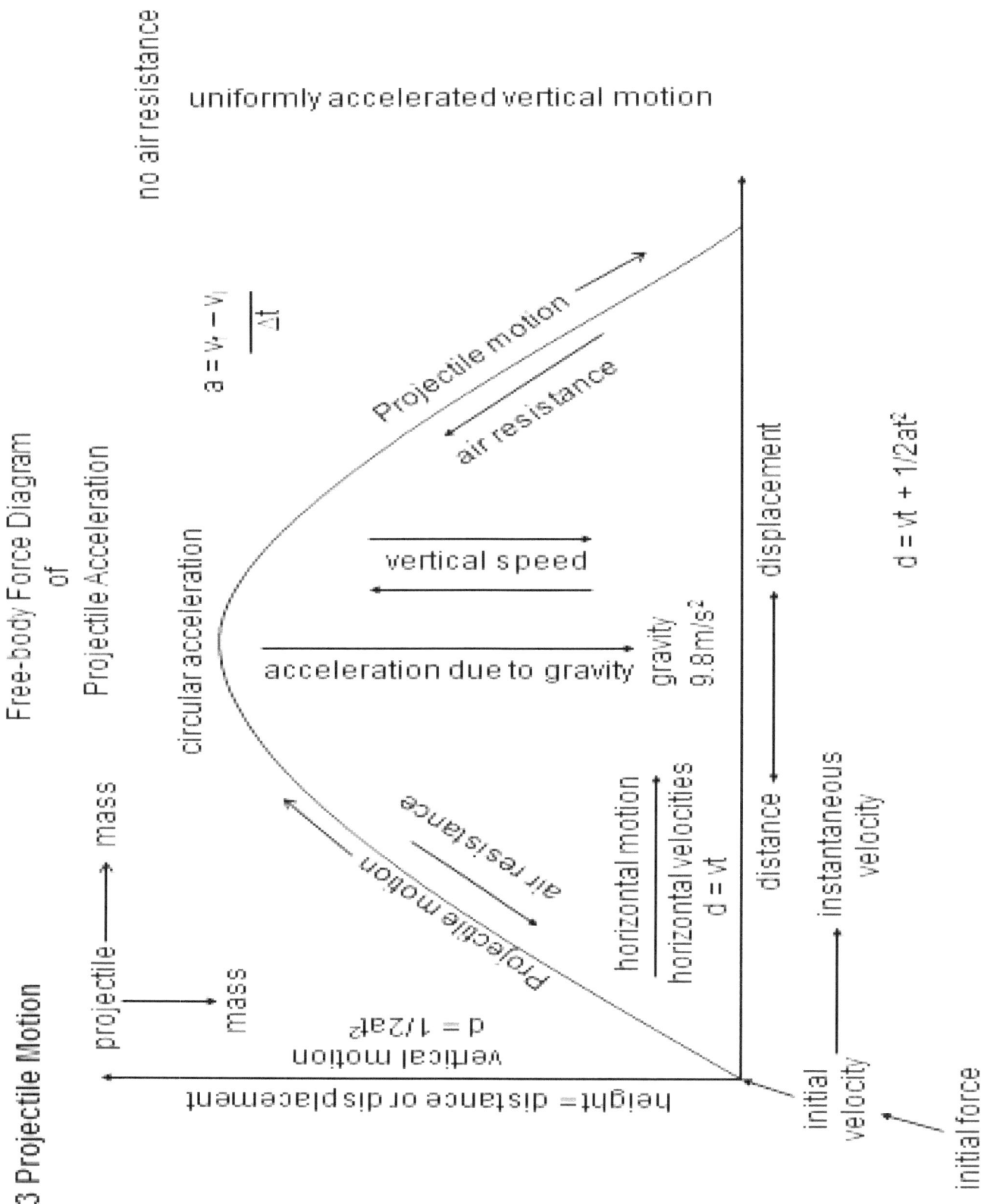

Picket Fence Free Fall

Directions:

You will need Vernier's **Picket Fence**, a **photogate** attached to an **interface** connected to a **computer** with **Logger Pro**, and a **clamp** and **ring stand** to secure the photogate. **Looking at the materials and lab we will be using, what are the safety precautions we should take to protect ourselves and materials during the investigation?**

1) Use the clamp to fix the photogate to the ring stand and move the ring stand to the table's edge.

2) In Logger Pro, open the Physics with Vernier folder and file #05 Picket Fence.

3) Click "Collect" to allow the photogate to turn on when you drop your picket fence through the photogate.

4) Then hold your picket fence over the photogate letting it hang down. Drop the photogate so that the alternating black and clear bands flow through the photogate. If one part of the picket fence does not go through in the drop, your reading will be off.

5) Look at your data, and make sure your slope for the velocity vs. time graph is a straight line; if it isn't, repeat #s 3 & 4.

6) If the graph's slope is straight, click the linear fit button and record the line's slope in the top data table on the next page; this is the acceleration of the picket fence dropping through the photogate.

7) Repeat steps 3-6 four more times.

8) Find the Minimum and Maximum values of the trials, and calculate the average by adding up all five slopes and dividing by 5. Write this value in the middle data table on the next page. It should be close to 9.81 m/s².

9) Find the precision by taking the lowest number of your average and 9.81 and dividing by the other, then multiply by 100, giving you the % accuracy. Write that down in the bottom data table.

Data Tables

Trial	1	2	3	4	5
Slope (m/s²)					

	Maximum	Minimum	Average
Acceleration (m/s²)			

Acceleration due to gravity, g	9.81 m/s²
Precision	%

Questions:

1) What is the shape of the position vs. time graph for each trial?

2) What is the shape of the velocity-time graph for each trial?

3) How close is your average acceleration compared to the 9.81 m/s²?

4) How would an acceleration vs. time graph look?

5) Would dropping your picket fence from a higher position change your acceleration results? Try it. Did it change the results; if so, how?

6) Would throwing the picket fence downward (but let it go before it enters the photogate) change your acceleration results? Try it. Did it change the results; if so, how?

7) Would throwing the picket fence upward (but let it go before it enters the photogate), letting it rise and fall, change your acceleration results? Try it. Did it change the results; if so, how?

8) Would adding air resistance (attaching some type of tail to the picket fence) change your acceleration results? Try it. Did it change the results; if so, how?

Simple Physics Investigations Seven Sides Publishing

Observing the Vertical Motion of a Kickball

Directions:

You will need a **small rubber kickball**, **two stopwatches** for each group, and lots of space outside. **Looking at the materials and lab we will be using, what are the safety precautions we should take to protect ourselves and materials during the investigation?**

1) **Hypothesis:** What do you think will be the relationship between the time it took the ball to rise and the time it took the ball to fall?
2) Two different students will have the stopwatches. One will start their stopwatch when someone kicks the ball and stop the stopwatch when the ball reaches its maximum height. The other student with a stopwatch will start the watch when the ball reaches its highest point and stop the watch when the ball hits the ground.
3) Have a student kick the ball, and the other two students measure the time going up and time coming down. Write this information in Data Table 1.
4) Average the times and answer the questions that follow.

Data Table 1

	Time Up	Time Down
1st Kick		
2nd Kick		
3rd Kick		
4th Kick		
5th Kick		
Average Time		

Questions:

1) What did you notice about the time it took for the ball to rise and the time it took for the ball to fall?

2) What forces were acting on the ball as it went up and down?

3) Why do you think the results came out as they did?

4) What could be some sources of error in the investigation?

5) How could this investigation be improved?

6) How would this investigation look different if there was no air resistance?

7) Write a formula that shows the time relationship between an object rising from the ground and the same object falling to the ground if there was no air friction during its flight up and down.

Measuring the Effects of Air Resistance

Directions:

You will need two **stopwatches** and two **pieces of paper**. **Looking at the materials and lab we will be using, what are the safety precautions we should take to protect ourselves and materials during the investigation?**

1) Take one piece of paper and wad it up into a ball. Leave the other paper flat.
2) Hold both pieces of paper at the same height and drop them at the same time. Make sure the flat paper is horizontal to the floor when you drop it. Have one person time the flat paper and the other person time the paper wadded up in a ball.
3) Write the times in Data Table 1 below.
4) Repeat the procedure in #s 2-3 two more times.
5) Find the averages of the times by adding the three times up and dividing by 3. Put this data in Data Table 1 below.

Data Table 1

Type of Paper	Trial 1 Time (s)	Trial 2 Time (s)	Trial 3 Time (s)	Average Time (s)
Flat				
Wadded				

Questions:

1) Which one accelerated faster?

2) Which one reached its terminal velocity first?

3) Explain why the two papers fell at different rates.

4) How does this relate to why we use parachutes when jumping out of planes?

Simple Physics Investigations

Air Resistance

Directions:

You will need a **scale**, five **coffee filters**, a **ring stand**, and a **motion detector** attached to an **interface** connected to a **computer** with **Logger Pro**. Looking at the materials and lab we will be using, what are the safety precautions we should take to protect ourselves and materials during the investigation?

1) Place the motion detector on the ring of the ring stand facing down.
2) In Logger Pro, open the folder Physics with Vernier and file #13 Air Resistance.
3) Find the coffee filter(s) mass and place that in Data Table 1 below.
4) Hold a coffee filter between .25m - .5m away from the motion detector.
5) Click "Collect" to begin data collection.
6) Drop the coffee filter so it will fall directly down under the motion detector.
7) The slope of the straight line will be the terminal velocity. Select the straight slope line and click the linear fit button to tell you the slope. Write that number in the data table below for one filter.
8) Repeat steps 3 through 7, adding another coffee filter for each trial, and put those numbers in Data Table 1 below.

Data Table 1

Number of Filters	Mass of the Filters (g)	Terminal Velocity (m/s)
1		
2		
3		
4		
5		

Questions:

1) How does the mass of the object affect terminal velocity?

2) How could surface area affect terminal velocity?

3) Does everyone need the same size parachute?

4) How big of a parachute would an elephant need compared to a human (Operation Dumbo Drop)?

Elevator Lab

Directions and Questions:

Take a **1 kg mass** into an **elevator** with a **digital scale. Looking at the materials and lab we will be using, what are the safety precautions we should be taking to protect ourselves and materials during the investigation?**

1) Place the 1 kg mass on the digital scale on the elevator floor.
2) What is the mass in the scale say before the Elevator moves?

3) Press a button in the elevator to move the elevator down. How does the reading on the scale change when the elevator starts to move down?

   (gets higher, gets lower)

4) How does the reading on the scale change when the elevator starts to slow down?

   (gets higher, gets lower)

5) Press a button in the elevator to move the elevator up. How does the reading on the scale change when the elevator starts to move up?

   (gets higher, gets lower)

6) How does the reading on the scale change when the elevator starts to slow down?

   (gets higher, gets lower)

7) When did the mass seem to have less weight?

8) Try to explain why.

9) When did the mass seem to have more weight?

10) Try to explain why.

Shoot and Drop

Directions:

You will need a **shoot and drop set-up** that simultaneously shoots a ball horizontally and drops another with no horizontal motion. **Looking at the lab and materials we will be using, what safety precautions should we take to protect ourselves and materials during this investigation?**

1) <u>Hypothesis</u>: Which ball do you think will hit the ground first?

2) Set up and fire the apparatus. Which ball hit the ground first?

3) How does horizontal motion affect vertical motion?

4) If you were to fire a gun horizontally to the ground and drop a bullet at the exact same; time, which bullet would hit the ground first.

Ball and Cart

Directions:

You will need a **ball and cannon cart** that shoots a ball straight up in the air while moving horizontally. **Looking at the lab and materials we will be using, what safety precautions should we take to protect ourselves and materials during this investigation?**

1) **Hypothesis**: Will the ball fall back into the cannon?

2) Push the cart and have it fire the ball into the air. Where did the ball land?

3) How does this show how horizontal motion affects vertical motion?

4) Why did your results come out the way they did?

5) How did inertia and momentum affect the results?

Projectile Motion

Directions:

You will need a **steel ball**, a **hot wheels ramp**, a **metric ruler**, **masking tape**, a **plumb**, and two **photogates** attached to an **interface** connected to a **computer** with **Logger Pro**. Looking at the materials and lab we will be using, what are the safety precautions we should take to protect ourselves and materials during the investigation?

1) Set up a hot wheels ramp and two photogates 10 cm apart on a lab table. The last photogate should be near the edge of the table; this should be set up so that if we place a marble on the ramp, it should roll down the ramp, then between the two photogates, then off the edge of the table.
2) Place a mark on the ramp where you will place the ball each time, allowing it to accelerate the same distance.
3) Make sure the photogate closest to the ramp (photogate 1) is plugged into DIG/SONIC 1 of the interface. And the photogate farthest from the ramp (photogate 2) is plugged into DIG/SONIC 2 of the interface. Ensure the interface is connected to a computer opened up to Logger Pro.
4) Open the folder Physics with Vernier and the file 08 Projectile Motion.
5) Make sure the distance between the centers of the photogates is exactly 10 cm.
6) On the Logger Pro display, enter the distance for Δs as .1 m.
7) Press "Collect" and place the steel ball on the mark you made on the ramp. Let the ball roll down the ramp (do not push), roll between the photogates, and catch it as it rolls off the table.
8) Place the ball back on the ramp's mark and let it roll down the ramp nine more times (you will have 2 minutes to do this).
9) Click "Stop" to stop the data collection.
10) Record the velocity for each trial in Data Table 1.
11) Find the maximum velocity, minimum velocity, and calculate the average velocity (add the ten velocities together and divide by 10). Put these numbers in Data Table 2.
12) Measure the height of the table in meters and write that down in Data Table 2.
13) Use the vertical projectile motion formula to calculate how long the ball will stay in the air as it drops off the table. $d = 1/2 \times 9.81 \times t^2$
14) Take this time found in #13 and calculate the steel ball's horizontal distance while in the air using the horizontal projectile motion equation and the average velocity we found with the photogates. $v = d \times t$

15) Use a plumb to find the spot on the floor under the table's edge.
16) Measure the distance you calculated in #14 from this spot under the table, away from the table; this will predict where the ball will land. Place a mark there. Write this distance in Data Table 2.
17) Repeat procedure #s 14-16 to find the minimum distance and maximum distance using the table's minimum and maximum velocities. Place those marks on the floor. Write these distances in the second data table.
18) Now roll your ball down the ramp and let it hit the floor. Did the ball land within these marks?

Data Table 1

Trial	Velocity (m/s)
1	
2	
3	
4	
5	
6	
7	
8	
9	
10	

Data Table 2

Maximum Velocity	m/s
Minimum Velocity	m/s
Average Velocity	m/s
Table Height	m
Predicted distance from the table	m
Minimum distance from the table	m
Maximum distance from the table	m
Actual impact point	m

Questions:

1) If we raise the table's height, how will this affect how far away from the table the steel ball hits?

   a. How would it change the vertical acceleration of the steel ball off the table?

2) If we lower the table's height, how will this affect how far away from the table the steel ball hits?

   a. How would it change the vertical acceleration of the steel ball off the table?

3) If we raise the ramp's height, how would this change the distance the steel ball lands away from the table?

   a. How would this change the vertical acceleration of the ball off the table?

4) What is the acceleration of the ball horizontally?

5) What is the acceleration of the ball vertically?

6) How does the vertical motion affect the horizontal motion?

7) How does the horizontal motion affect the vertical motion?

Virtual Investigations that go with Projectiles

ExploreLearning.com:

    Free-Fall Laboratory Gizmo

    Feed the Monkey Gizmo

    Golf Range Gizmo

PhET.colorado.edu:

    Projectile Motion

Physicsclassroom.com:

    Physics Interactives:

        Vectors and Projectiles

        Projectile Simulator

        Turd the Target

        Turd the Target 2

        Monkey and the Zookeeper

    Concept Builders:

        Vectors and Projectiles

        Free Fall

        Up and Down

        Which One Doesn't Belong? Projectile Motion

        Trajectory – Horizontally Launched Projectiles

        Trajectory – Angle Launched Projectiles

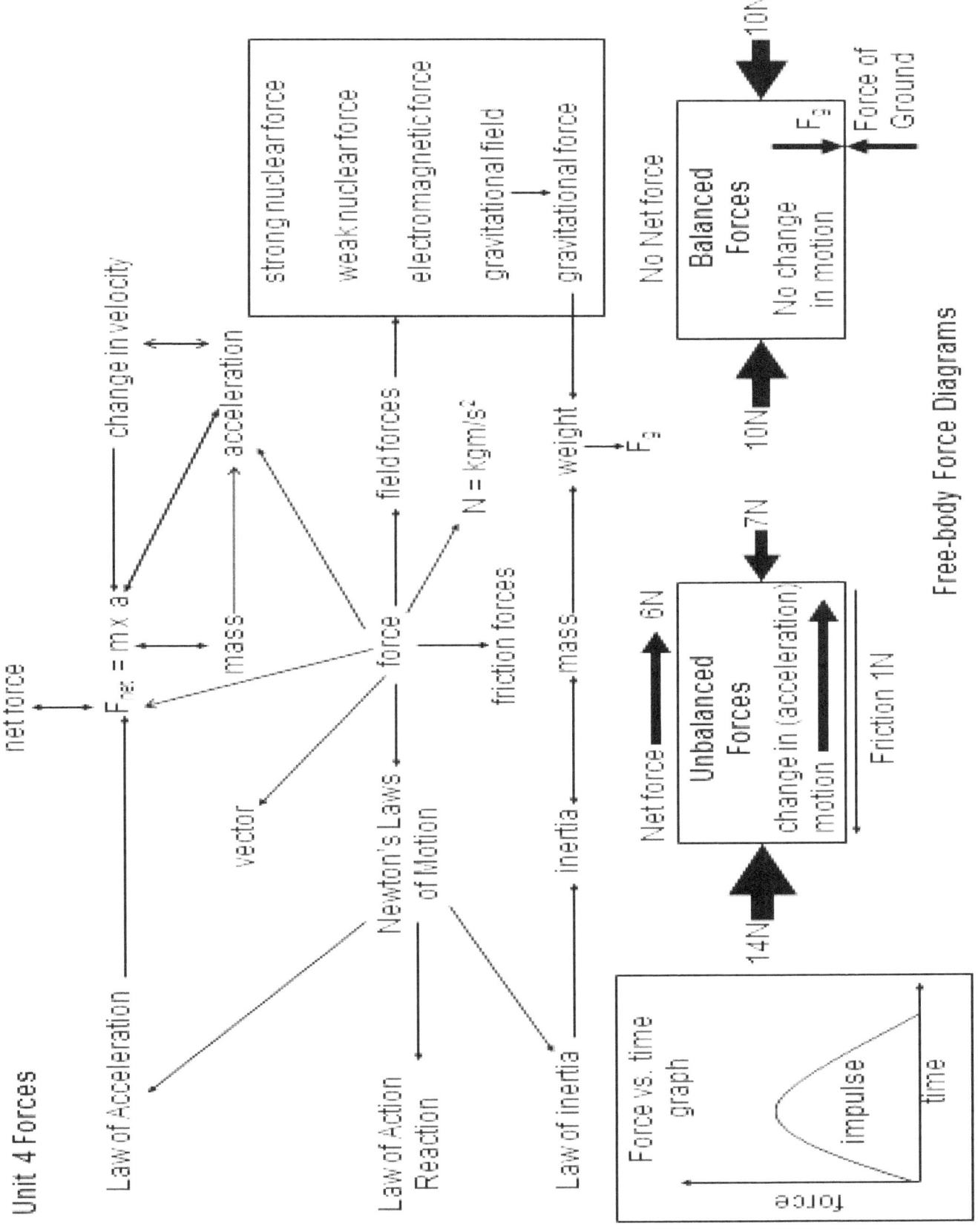

The Human Table

Directions:

You will need four **chairs** and four **people** about the same height. **Looking at the materials and lab we will be using, what are the safety precautions we should take to protect ourselves and materials during the investigation?**

1) Set the four chairs facing each other in a square a couple of feet apart. Distances can vary depending on the size of the people you are using.
2) Have the students sit on the chairs with their legs on the right-hand side of the chairs.
3) The student should then be able to lean back and place their heads on the legs just above the knees of the person behind them. Adjust the chairs to make sure each will be able to do this before they lean back.
4) Have the students lean back and place their heads on the person's legs behind them. Have each person lift their butts off the chairs and pull the chairs out away from the student square. You should now have a human table.
5) The students should be able to hold this for a few minutes, discuss the questions with the rest of the class, then put the chairs back under the butts of the students so they can sit up.

Questions:

1) What was holding the students up?

2) What are the forces, and how are they involved in this system?

3) Was it balanced or unbalanced? Explain why.

Simple Physics Investigations — Seven Sides Publishing

Balancing Forks

Directions:

You will need two **forks**, a **toothpick**, and a **glass**. **Looking at the materials we will be using, what are the safety precautions we should take to protect ourselves and materials during the investigation?**

1) Take the two forks and force them together at the prong end.
2) Take a toothpick and place it on the rim of the glass; simultaneously, balance the two forks on the toothpick with the forks' handles going around the sides of the glass.
3) Discuss with the class and teacher why you think this is happening.

Questions:

1) What do you think is causing the forks to be balanced on the toothpick without falling?

2) What forces are acting on this system?

   a. How are they involved in this system?

3) Draw a force diagram of this system below showing the forces and how they are balancing the system.

Weight isn't Mass Lab

Directions:

You will need a **set of varying masses** from 1 kg to 1 g. You will also need a **force sensor** attached to an **interface** connected to a **computer** with **Logger Pro**. Looking at the materials and lab we will be using, what are the safety precautions we should take to protect ourselves and materials during the investigation?

1) Hold the force sensor with the hook hanging down and hang each mass on the hook to measure force in Newtons. Write this data in Data Table 1.
2) Once Data Table 1 is complete, take this data and plot it on Graph 1 below.

Data Table 1

Mass (kg)	Force (N)

Graph 1: Mass vs. Force

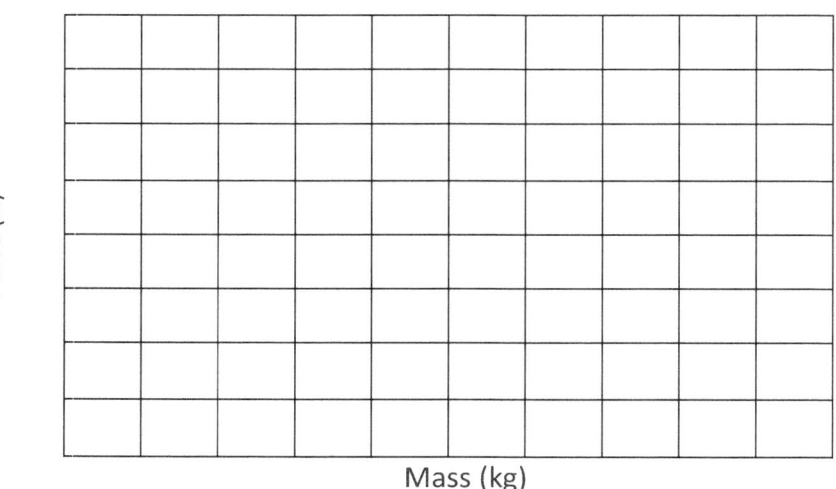

Questions:

1) Calculate the slope of the graph. Show your work.

2) What is the significance (meaning) of your slope?

3) Write the mathematical model (equation) of your graph?

4) Weight is the force of gravity acting on a mass. The force of the object pushing down is equal to the weight. Force can be calculated as the acceleration due to gravity times the mass (F = m x a). How does this relate to the equation you wrote in #3?

5) If you were to conduct this same experiment on the moon instead of the Earth, how would your graph's results be different from Earth's?

6) If the mass is doubled, what would happen to its weight?

Comparing Friction Lab

Directions:

You will need an **ice cube, rock, eraser, wooden block, aluminum foil**, and a **tray. Looking at the materials and lab we will be using, what are the safety precautions we should take to protect ourselves and materials during the investigation?**

1) Position each object on your tray.
2) Slowly lift one end of the tray and stop when an object slides.
3) Measure the height of the end of the tray you raised.
4) Keep doing this until you have measurements for all of your objects.
5) Put all your data in Data Table 1 below.

Data Table 1

Object	Height at which the object slid (cm)
Ice Cube	
Rock	
Eraser	
Wooden block	
Aluminum foil	

Questions:

1) Why did the objects slide off at different heights?

2) What type of friction did each object overcome to start sliding?

3) What type of friction does the object have as it slides down the tray?

4) Which type of friction do you think is greater?

5) Which direction is friction from the movement?

Friction Lab

Directions:

You will need a **scale**, **100 g, 200 g, 500 g masses**, and a **wooden block with a hook** attached. You will want to perform this experiment on three **different types of surfaces**. You will also want to have a **force sensor** attached to an **interface** connected to a **computer** with **Logger Pro**. You could use a **spring scale** instead of the force sensor, but it will not be as accurate. **Looking at the materials and lab we will be using, what are the safety precautions we should take to protect ourselves and materials during the investigation?**

1) Measure the mass of the block and record it in the data on page 68. Calculate its weight by multiplying it by 9.81 and writing it in the data on page 68.
2) Measure the block's surface's length and width in contact with the desktop (for both wide and skinny sides). Write this in the Data Table 1.
3) Attach the force sensor to the hook on the block.
4) Set the Logger Pro to collect data for 5 seconds.

Part 1 Experiment: Relationship between the surface area and the friction force.

5) Set the bock with the wide side down. As you start moving the force sensor, turn on the "Collect" data button and pull evenly for five seconds. Click on the STAT button on the toolbar in Logger Pro to find the average force from the procedure's data and record it in Data Table 1.
6) Repeat # 5 for two more trials.
7) Find the average of the three trials (add them together and divide by 3). Write this in Data Table 1.
8) Calculate the block's wide side friction coefficient by taking the average force and dividing it by the weight force. Write this in Data Table 1.
9) Repeat the procedures for #5-8 for the skinny side of the block. Write that information also in Data Table 1.

Part 2 Experiment: Relationship between the surface type and the kinetic and static friction forces.

10) Record the three types of surfaces you will pull your block across in Data Table 2.
11) Put your block on the first surface wide side down.
12) Repeat the procedures for #5-8 for each surface type and write your data in Data Table 2. The static friction (friction when the block is not moving) is the maximum value for

the data collected. The average value for the data collected is the kinetic friction (friction when the block moves).

Part 3 Experiment: Relationship between the normal force and the frictional force.

13) Put your block on a wide surface side down, add 100 g mass to the block and repeat the procedures in #s 5-8. Write data in Data Table 3.
14) Take off the 100 g mass, add a 200 g mass to the block, and repeat the procedure for #s 5-8. Write data in Data Table 3.
15) Take off the 200 g mass, add a 500 g mass to the block, and repeat the procedure for #s 5-8. Write data in Data Table 3.

Part 4 Experiment: Relationship between the type of shoe and the frictional force.

16) Have a person in the group take off their shoe and add a 500 g mass to the shoe's inside. Hook the force sensor to the shoe, repeat the procedures for #s 5-8, and write the data in Data Table 4.
17) Repeat #16 for each person in the group. Write their data in Data Table 4.

Data:

Mass of the block _____ kg

Weight of the block or normal force (m x 9.81m/s²) _____ N

Data Table 1

Side	Surface area	Trial 1 Force (N)	Trial 2 Force (N)	Trial 3 Force (N)	Average Force (N)	µ=Average force divided by Weight
Wide						
Skinny						

Data Table 2

Type of surface		Trial 1 Force (N)	Trial 2 Force (N)	Trial 3 Force (N)	Average Force (N)	µ=Average force divided by Weight
1	Static					
1	Kinetic					
2	Static					
2	Kinetic					
3	Static					
3	Kinetic					

Data Table 3

Total mass (kg)	Normal Force (N)	Trial 1 Force (N)	Trial 2 Force (N)	Trial 3 Force (N)	Average Force (N)	µ=Average force divided by Weight

Data Table 4

Type of shoe	Normal Force	Trial 1 Force (N)	Trial 2 Force (N)	Trial 3 Force (N)	Average Force (N)	µ=Average force divided by Weight

Questions:

1) Did the size of the surface area affect the frictional force? Why or why not?

2) Explain any difference between the values for the static and sliding friction coefficients.

3) Which surface seemed to have the highest coefficient of friction?

4) What was the relationship between weight and the coefficient of friction?

5) Which shoe in the group had the largest coefficient of friction?

6) Which shoe had the smallest coefficient of friction?

7) Which shoe would you want to wear when playing basketball?

8) Which shoe would be best used for sliding across the floor while dancing?

9) Why do hiking boots have rubber soles and deep tread?

10) Why do racing tires on cars have no tread while normal tires do have tread?

11) If the objects were being pulled at a constant velocity in the experiment, were the pulling force and friction force balanced? Show a force diagram explaining your answer.

Simple Physics Investigations   Seven Sides Publishing

Newton's Relay Race

Directions and Questions:

You will need a **broom**, a **bowling ball**, a **basketball**, and a kid's **rubber ball**. **Looking at the materials and lab we will be using, what are the safety precautions we should take to protect ourselves and materials during the investigation?**

Accelerating an object from rest: we will be observing <u>inertia</u> – resistance to change motion (Newton's 1st Law)

1) Place the bowling ball on the floor. Push the ball with a broom in a sweeping motion to cause the bowling ball to accelerate. When does the bowling ball accelerate?

2) How easy was it to accelerate?

3) When does it move at a constant speed?

4) Place the kid's rubber ball on the floor and push this ball the same way you did with the bowling ball. When does the ball accelerate?

5) How easy was it to accelerate?

6) When does the ball move at a constant speed?

7) Which ball had the most inertia?

8) Draw a force diagram of the action of accelerating the ball.

Stopping an object: observing inertia (Newton's 1st Law) and momentum – resistance to stop motion

9) Get the bowling ball moving, then stop it with the broom. Do the same with the kid's rubber ball. Which ball was harder to stop?

10) Which ball had the most momentum?

11) When sitting still, which object has the most inertia?

12) When sitting still, which object has the most momentum?

13) Draw a force diagram of the action of stopping a moving ball.

Turning an object 180 degrees: observing inertia (Newton's 1st Law) and momentum

14) Get the bowling ball and start moving it, then stop it and turn it 180 degrees. Do the same for the kid's rubber ball. Which ball was easier to change direction?

15) Which ball had the most momentum?

Applying a constant force on the ball: observing force and acceleration (Newton's 2nd Law)

16) Go to a long hallway and make sure it is clear. Get the bowling ball moving by pushing it with the broom; see what happens if you try to put a constant force on the ball. Do the same for the kid's rubber ball. What happened to the speed of both balls?

17) Could you keep doing it?

18) Which ball could you apply the force for the longest amount of time? Why?

19) Which ball accelerated the fastest?

20) How could you make the balls move at a constant velocity?

Relay Race: observing inertia (Newton's 1st and 2nd Laws) and momentum

21) Make a course that changes direction several times (I have made my student go around the center demo table in my room) to push the different balls around in a race. Divide up into three equal teams, each with a broom. One team will push a bowling ball with a broom, one team will push a basketball with a broom, and one group will push the kid's rubber ball with the broom. Predict which team will finish the course first with all its team members.

22) Have the students do a relay race to move the balls through the course (be very careful of the bowling ball and make sure kids know to move out of its way if it comes at them). Which team won?

23) Why were they able to win?

24) Which team came in last? Why?

Observing Inertia Newton's First Law of Motion

Directions:

You will need a **toy car that winds up** when you pull it back and then moves forward when you let it go. You will also need a **penny** and a **rubber band. Looking at the materials and lab we will be using, what are the safety precautions we should take to protect ourselves and materials during the investigation?**

1) Place the penny on the car. Pull the car back to wind up the car. Let it move forward and have it crash into something.
2) What happens to the penny?

3) Place a rubber band around the car and use it to fix the penny to the car.
4) Repeat the procedure in # 1
5) What happens to the penny?

Questions:

1) Why did the penny do what it did in the first crash?

2) Explain the forces acting in both crashes.

3) What does the penny represent?

4) What does the rubber band represent?

5) Why should we wear seat belts when we get in a car?

6) State Newton's First Law of Motion.

Simple Physics Investigations   Seven Sides Publishing

Inertia Lab Stations

Equipment and Safety:

You will need two **ping pong balls**, a **ping pong paddle**, two **tennis balls**, an **index card**, a **cup**, and a **penny**. **Looking at the materials and lab we will be using, what are the safety precautions we should take to protect ourselves and materials during the investigation?**

Remember that **inertia** is the resistance of objects to change their motion. Objects in motion tend to stay in motion, and objects at rest tend to stay at rest; unless an outside force acts on them. The larger the mass, the more inertia it has. Think of this as you go through the three lab stations.

Station 1

Take a ping pong ball and use the paddle to bounce it off the door or wall a few times. Now take a tennis ball and use the paddle to bounce it off the door or wall a few times.

1) Which ball seems to have more inertia?

2) How can you tell?

3) Why does that ball have more inertia?

4) Which ball can you get to move faster off the paddle? Why?

Station 2

Place an index card on the opening of the cup. Now place a penny on top of the index card. With one hand, hold the cup; with the other hand, grab the index card and quickly pull it away horizontally.

1) What happened to the penny?

2) Why did it not move horizontally with the card?

3) What forces acted on the penny after the card was pulled away?

Station 3

1) Roll a ping pong ball at a stationary tennis ball, have them collide, and describe what happens.

2) Now roll the tennis ball at a stationary ping pong ball, have them collide, and describe what happens.

3) Use your knowledge of the Law of Inertia to explain why the results were not the same.

Newton's Second Law

Directions:

You will need a **scale**, a **cart**, a long **rubber band**, a **dual-range force sensor**, and **low g accelerometer**, and an **interface** connected to a **computer** with **Logger Pro**. **Looking at the materials and lab we will be using, what are the safety precautions we should take to protect ourselves and materials during the investigation?**

1) Stack the dual-range force sensor on top of the cart and the accelerometer on the force sensor. Have them face the same direction as the wheels will be moving, so the force and acceleration will be measured in the same direction. Tightly wrap the rubber band around the system, holding it all together.

2) Connect the dual-range force sensor to channel 1 on the interface. Connect the low g accelerometer to channel 2.

3) Open Physics with Vernier folder and file #09 Newton's Second Law.

4) Click "Collect" to collect data. Holding the hook of the dual-range force sensor, roll the cart back and forth in the direction the wheels move. Vary the forces, both small and large.

5) What is the shape of the force vs. time graph?

6) What is the shape of the acceleration vs. time graph?

7) Click the examination button. Move the mouse across one of the graphs. When the force is at its maximum, what is the acceleration? (maximum or minimum)

8) Click on the Force vs. Acceleration graph and click the linear fit button. Record the slope of this line in Data Table 1.

9) Find the mass of the cart and sensors. Write this in Data Table 1.
10) Add/fix a 500 g mass to the cart, repeat the procedure for #4 and 8, and record the slope of the force vs. acceleration graph in Data Table 1.
11) Find the mass of everything in the cart now. Write that in Data Table 1.

Data Table 1

Cart	Slope of graph	Mass (kg)
Cart and Sensors		kg
Cart, sensors, and 500 g		kg

Questions:

1) How are the Force vs. Time and Acceleration vs. Time graphs similar for the two trials?

2) How are they different?

3) Compare the slope of the Force vs. Acceleration graph and the mass. What does the slope represent?

4) Write a general formula for the three variables: force, mass, and acceleration.

5) What is the unit for the slope of the Force vs. Acceleration graph?

6) What is the relationship between force and acceleration in the equation?

7) What is the relationship between the mass and acceleration in the equation?

8) How could this information help you run away from a rhinoceros that is chasing you?

9) If you have a bowling ball and a baseball, each is suspended by a separate rope, and try to hit each with a baseball bat – Which ball will have the biggest change in motion? Explain why.

Fan Cart Lab

Directions:

You will need a **scale**, a **mass**, a **cart**, a **fan** to fix to the cart, a **motion detector** attached to an **interface** connected to a **computer** with **Logger Pro**, and a **Dynamics System** from Vernier. **Looking at the materials and lab we will be using, what are the safety precautions we should take to protect ourselves and materials during the investigation?**

1) Use the scale to measure the mass of the cart. Write this in Data Table 1.
2) Position the motion detector at one end of the track. Position the cart in front of the motion detector so it will move away from the motion sensor.
3) Start the fan. Click the "Collect" button. Allow the cart to accelerate down the track. Click "Stop." Grab the cart before it falls off the track.
4) Highlight the line on the Velocity vs. time graph. Click the "Fit" button and choose linear fit. The slope of the line is the acceleration of the cart. Record the acceleration in Data Table 1.
5) Repeat #s 3-4 two more times and put the accelerations in Data Table 1.
6) Calculate the average acceleration by adding the three values and dividing by 3.
7) Add a mass to the fan cart. Record the amount of the mass in the data below.
8) Repeat the procedures in #s 3-6. Put this data in Data Table 1.

Data:

Mass of the cart and fan _____ kg   Amount of mass added _____ kg

Data Table 1

Cart	Mass (kg)	Trial 1 Acceleration	Trial 2 Acceleration	Trial 3 Acceleration	Average Acceleration
Cart + Fan					
Cart + Fan + Mass					

Questions:

1) Using the formula F = m x a, calculate the force of the fan on the cart. What is that force?

2) Do the same for the cart with the mass on it. What is that force?

3) Compare the answers for #s 1 and 2.

4) Why did they have different accelerations then?

5) What may be a source of error not figured in the equation?

Newton's Third Law

Directions and Questions:

You will need a **rubber band** and two **dual-range force sensors** attached to an **interface** connected to a **computer** with **Logger Pro**. Looking at the materials and lab we will be using, what are the safety precautions we should take to protect ourselves and materials during the investigation?

1) Hold a rubber band between your right hand and your left hand. Pull with your left hand. Does your right hand experience a force?

2) Does your right hand apply a force to the rubber band?

3) What direction is this force compared to the left hand?

4) Pull harder with your left hand. Does this change any force applied by your right hand?

5) How is your left hand's force, transmitted by the rubber band, related to the force applied by your right hand?

6) Write a rule in words for this force relationship.

7) Place the two dual-range force sensors opposite each other with their hooks facing each other. Attach the rubber band between them.

8) In Logger Pro, open the folder Physics with Vernier and file #11 Newton's 3rd Law.

9) Press "Collect" and have one person pull on one sensor and another person pull on the other back and forth, stretching and shortening the rubber band. What does the graph show about the magnitude of both forces?

10) What does the graph show about the direction of the two forces?

11) Is there any way to pull on your force sensor without your partner's force sensor pulling back while keeping tension on the rubber band?

12) Fasten one force sensor to your lab table and repeat the experiment. Does the lab table pull back?

13) Connect the two force sensors together with just their hooks instead of the rubber band. How do the results change?

14) State Newton's Third Law of motion:

15) How does this lab show Newton's Third Law of Motion?

16) If you are driving down the street and a bug splatters on your windshield – What is greater: the force of the bug on the windshield or the force of the windshield on the bug? Explain why.

Water Bottle Rockets

Directions and Questions:

You will need a **water bottle rocket launcher**, a **2-liter bottle**, and an **air pump** with a **pressure gauge. Looking at the materials and lab we will be using, what are the safety precautions we should take to protect ourselves and materials during the investigation?**

1) Fill the water bottle half full with water.
2) Angle the launcher straight up at a 90-degree angle to the ground (gives the rocket its highest distance to travel in the air).
3) Pump 20 pounds of pressure into the rocket. Start a stopwatch when you launch the rocket and stop it when it reaches the highest point in the air. What was the time?

4) Calculate the initial velocity of your rocket by using your Formula Chart to find the formula that works. Show all work, including units and formulas.

5) Using this data, calculate the approximate distance your rocket went into the air. Use your Formula Chart to find the formula. Show all work, including units and formulas.

6) What do you think will happen to the time in the air and the launch's height if we double the pressure?

7) Pump 40 pounds of pressure into the rocket. Start a stopwatch when you launch the rocket and stop it when it reaches the highest point in the air. What is the time?

8) Calculate the initial velocity of the rocket by using your formula chart. Make sure to show all work, including units and formulas.

9) Using this data, calculate the approximate distance your rocket went into the air. Use your Formula Chart to find the formula. Show all work, including units and formulas.

10) How did changing the pressure affect the flight of your rocket?

11) What other variables could we change to affect the rocket's height if you were to launch again?

12) How do you think changing the rocket's mass will affect the force of gravity?

13) How would it affect the inertia?

14) How is Newton's 1st Law of Motion affect the rocket launch?

15) How does Newton's 2nd Law of Motion affect the rocket launch?

16) Draw a force diagram to show the variables affected in Newton's 2nd Law of Motion.

17) How is Newton's 3rd Law of Motion seen in the launch?

18) Draw a force diagram to show how the launch shows Newton's 3rd Law of Motion.

Virtual Investigations that go with Force

ExploreLearning.com:

    Weight and Mass Gizmo

    Free Fall Tower Gizmo

    Inclined Plane Sliding Objects Gizmo

    Fan Cart Physics Gizmo

    Crumple Zones Gizmo

    Force and Fan Carts Gizmo

    Atwood Machine Gizmo

    Determining a Spring Constant Gizmo

    Polarity and Intermolecular Forces Gizmo

    Archimedes' Principle

PhET.colorado.edu:

    Balancing Act

    Forces and Motion

    Forces and Motion Basics

    Forces in one Dimension

    Friction

    Lunar Lander

    Masses and Springs

    Masses and Springs Basics

    Ramp: Forces and Motion

    The Ramp

    Torque

Hook's Law

Physicsclassroom.com:

Physics Interactives:

Newton's Laws

Force

Free Body Diagrams

Rocket Sled

Skydiving

Elevator Ride

Atwood's Machine

Concept Builders:

Newton's Laws

Balanced vs. Unbalanced Forces

Force and Motion

Change of State

Recognizing Forces

Match That Free-Body Diagram

Normal Force Card Sort

Which One Doesn't Belong? Force and Motion

Newton's Second Law-Equations as Guides to Thinking

Net Force (and Acceleration) Ranking Tasks

Air Resistance and Skydiving

Fnet = m x a

Solve It! with Newton's Second Law

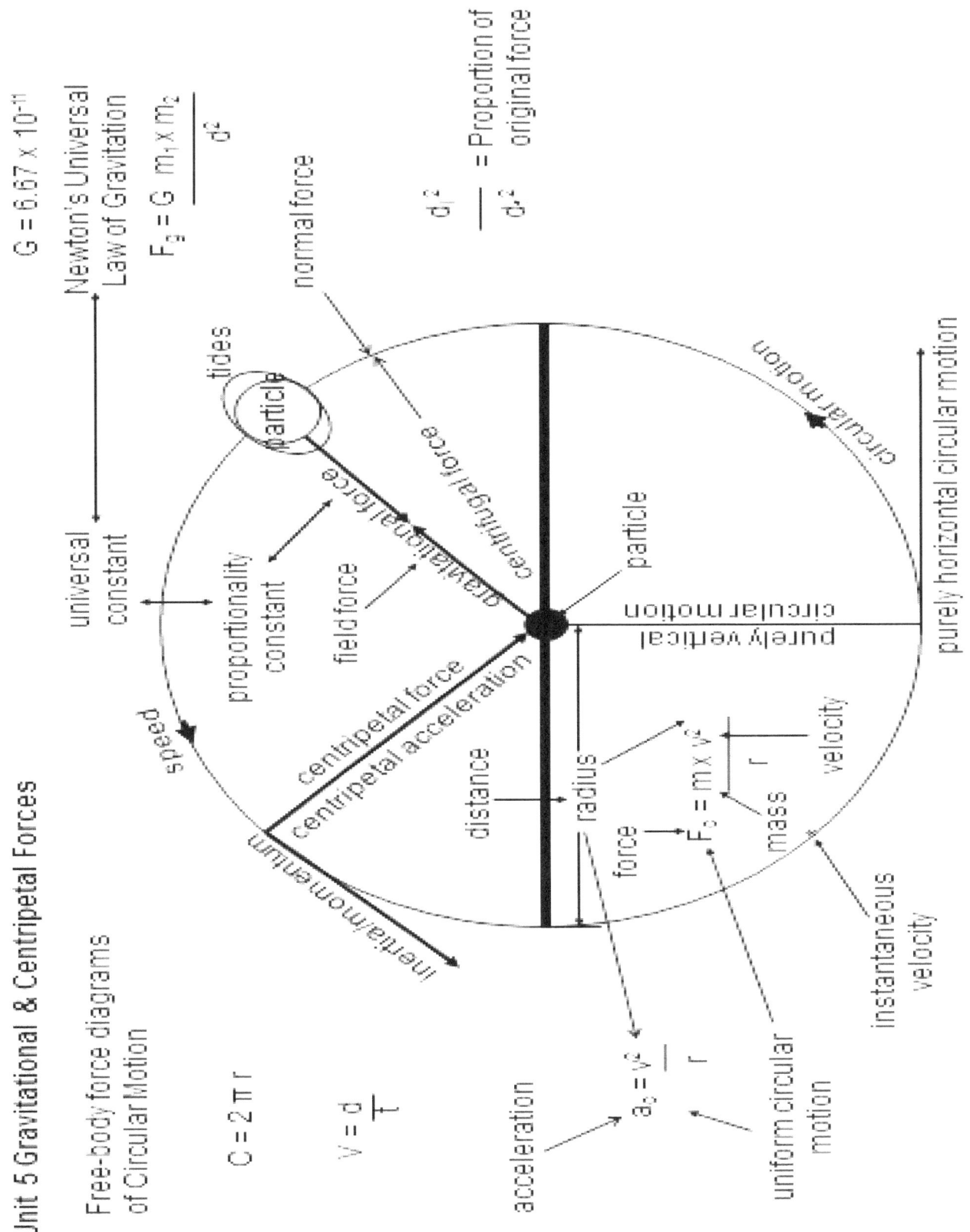

Bending of Space-time

Directions and Questions:

You will need the largest **Embroidery Hoop/Cross Stich Hoops** you can find. One for each group in your class. You will need to stretch some **elastic fabric** across and fix it into the hoops; this will represent the fabric of space-time. You will also need some **different weighted balls** and a **marble** to represent objects in the universe. **Looking at the materials and lab we will be using, what are the safety precautions we should take to protect ourselves and materials during the investigation?**

1) Have one student hold the hoop parallel to the ground. Have another student place the lightest ball inside the hoop in the fabric. What do you see happen to the fabric?

2) Now take the first ball off and place a heavier ball on the fabric. How was this different from the first ball?

3) How does this model show how objects bend the space-time continuum?

4) Roll a marble into your model with a heavy ball in the middle. What does it do to the larger object?

5) How does this show how gravity works in our universe?

6) Does this show gravity is a push or a pull? Explain why. Hint think of the ball in the center as like the Earth.

7) How would the fabric bend if you had an infinite mass in the center?

8) What would you make?

9) How is this model we used not accurate?

The Push of Gravity on Earth

Directions and Questions:

Use the **internet** to find the masses of the **Earth** and its **moon**, the distance from the moon to the Earth at their closest distance and their farthest distance. Use Newton's Universal Law of Gravitation to help you calculate the change in gravitational force between them. Make sure you show your work in the formulas you used.

1) Use Newton's Universal Law of Gravitation again to calculate the change in gravitational force between the **Sun** and the **Earth**.

2) Which object has a bigger effect on the tides of the Earth's oceans, the Sun or the moon? Give evidence and explain why.

3) What are the variables that affect the force of gravity between objects?

4) Calculate the force of gravity between your body and the Earth.

5) How much would the force of gravity change if you were on the surface of the moon?

6) If you were on the moon, which object would have a bigger gravitational force on you, the Earth or the moon? Give evidence and explain why.

7) Why does the title of this investigation say gravity has a push instead of a pull?

8) Calculate the acceleration due to gravity if you were on the surface of the moon.

Observing Centripetal Force

Directions:

Put some **water** in a **bucket** with a strong handle and get a **penny** and a **wire coat hanger**. **Looking at the materials and lab we will be using, what are the safety precautions we should take to protect ourselves and materials during the investigation?**

1) Stand away from others. The best place to go is outside.
2) Swing the bucket in a circle makes the bucket go upside down at one point. Make sure it keeps moving during the observation.

Questions:

1) Why did you not get wet?

2) Which direction does the water want to go?

3) Where is centripetal acceleration?

4) Where is centripetal force?

5) Where is centrifugal force?

6) Where is the normal force?

7) What force was exerted by your arm on the bucket?

8) What force was exerted by the bottom of the bucket?

9) What force was exerted by the water in the bucket?

10) What would happen if you swung the bucket slowly?

11) How does this investigation show how and why the Earth orbits the Sun, and the Moon orbits the Earth?

12) Take the wire coat hanger and bend the triangle's bottom side to make a square so you can spin the hanger around your finger there. You may have to file down the end of the hook so you can balance a penny on it. Once the penny is balanced (works best tails side down), gently spin the hanger around your finger. The penny should stay balanced on the hanger if you do it right. Explain why this happens.

Centripetal Force Under Glass

Directions:

You will need a **glass** or a **see-through plastic cup** and a **marble** on a tabletop. **Looking at the materials and lab we will be using, what are the safety precautions we should take to protect ourselves and materials during the investigation?**

1) Place a marble in a glass/plastic cup. Move the glass/cup in a circular motion causing the marble to spin around the inside of the glass/cup off the table's surface.
2) Discuss what you are seeing with the class and your teacher.

Questions:

1) Explain why the marble can stay off the surface of the table.

2) Where is the centripetal force in this lab?

3) Where is the centrifugal force in this lab?

4) What happens when you stop the movement of your glass/cup?

Uniform Circular Motion Lab

Directions:

You will need a **meter stick**, **hanging masses**, at least a meter of **string**, a **rubber stopper with a hole**, a **stopwatch**, and a **PVC pipe**. **Looking at the materials and lab we will be using, what are the safety precautions we should take to protect ourselves and materials during the investigation?**

Part 1: Relationship between the radius and average speed (force stays constant)

1) Tie a rubber stopper at one end of the string with the string going through a PVC pipe and 500 g mass on the end of the string hanging down.
2) Measure 20 cm of the string from PVC to the rubber stopper.
3) Spin the stopper to where you have the stopper moving in a constant rhythm. Record the time it takes for 20 complete revolutions (cycles).
4) Repeat steps 1-3 until you have data for all radii in Data Table 1. Make sure you try to keep the same rhythm for each distance.

Data Table 1

Radius (m)	Time for 20 cycles	Period average time per cycle	Circumference ($2\pi r$)	Average Speed (m/s)
.20 m	s	s	m	m/s
.40 m	s	s	m	m/s
.60 m	s	s	m	m/s
.80 m	s	s	m	m/s

Questions Part 1:

1) Is the stopper accelerating as it is spinning in a circle? Why or why not?

2) Draw a force diagram for the stopper spinning in a circle.

3) Say, while spinning the stopper, the string suddenly breaks. Draw a force diagram for the stopper right after the string breaks. Make sure you draw the path of the stopper after the string breaks.

4) Which radius had the fastest speed?

5) How does this information work for golf in choosing which club to hit?

Part 2: Relationship between force and speed

Directions:

1) Set your radius to 40 cm.
2) Increase the force applied to your stopper by adding 100 g to the hanging mass, making a total of 600 g.
3) Calculate the force in Newtons and put that in Data Table 2.
4) Spin the stopper to where you have the stopper moving in a constant rhythm. Record the time it takes for 20 complete revolutions (cycles).
5) Add another 100 g to the hanging mass for a total of 700 g and repeat the procedure in #s 2-4. Record the data in Data Table 2.
6) Add another 100 g to the hanging mass for a total of 800 g and repeat the procedure in #s 2-4. Record the data in Data Table 2.

Data Table 2

Force	Time for 20 cycles (s)	Period: time/20 (s)	Distance traveled (2πr)	Average speed (m/s)
N	s	s	m	m/s
N	s	s	m	m/s
N	s	s	m	m/s

Questions Part 2:

1) How does the force relate to speed in uniform circular motion?

2) How does this apply to a car turning a corner?

3) Since we have difficulty increasing the friction force when going around a corner, what should we do when safely turning a corner in a vehicle?

4) What is another way to turn that corner at a higher speed?

Simulating the Orbit of a Planet and Sun

Directions and Questions:

You will need a small **light ball** tied to a **string** and a **heavier ball** tied to a string. Your head will simulate the sun, and the ball will simulate a planet. **Looking at the materials and lab we will be using, what are the safety precautions we should take to protect ourselves and materials during the investigation?**

1) Hold the end of the string up near your head and swing the ball around your head. Notice your head moves back and forth, wobbling with the motion of the ball; this is similar to how a sun wobbles to the orbit of a planet. Why do you think the sun wobbles?

2) Make sure you orbit the small ball, then the heavier ball around your head. Which one caused your head to wobble more?

3) What do you think causes the size of the wobble of a sun?

4) When looking at a star, what do you think a physicist could tell from the wobble of a star?

Observing Forces in Orbits

Directions:

Tie a **string** to a **tennis ball**, and get a **penny** and a **wire coat hanger**. Looking at the materials and lab we will be using, what are the safety precautions we should take to protect ourselves and materials during the investigation?

Stand away from others, tightly hold one end of the string and spin the tennis ball around you above your head.

Questions:

1) Which direction does the ball want to go?

2) How can you tell?

3) How does this investigation show how and why the Earth orbits the Sun, and the Moon orbits the Earth?

4) Which object represents gravity?

5) Which object represents the sun?

6) Which object represents the Earth going around the sun?

7) Which object represents the moon going around the Earth?

8) What do you think causes the bubble of high tide between the Earth and the moon?

9) How does the moon's inertia affect the moon's path of motion relative to the Earth?

10) Take the wire coat hanger and bend the triangle's bottom side to make a square so you can spin the hanger around your finger there. You may have to file down the end of the hook so you can balance a penny on it. Once the penny is balanced (works best tails side down), gently spin the hanger around your finger. The penny should stay balanced on the hanger if you do it right. Explain why this happens.

11) Which object is being orbited?

12) Which object is doing the orbiting?

13) Which object represents the force of gravity?

14) How does this model show how gravity works on orbiting objects?

15) How is this model not accurately showing how gravity works on objects in orbit?

Virtual Investigations that go with Gravitational and Centripetal Forces

ExploreLearning.com:

    Gravitational Force Gizmo

    Uniform Circular Motion Gizmo

    Gravity Pitch Gizmo

    Orbital Laws Kepler's Laws Gizmo

    Moment of Inertia Gizmo

    Solar System Explorer Gizmo

    Tides Gizmo

    Ocean Tides Gizmo

PhET.colorado.edu:

    Gravity and Orbits

    Gravity Force Lab

    Gravity Force Lab Basics

    Ladybug Revolution

    My Solar System

    Motion 2D

    Torque

Physicsclassroom.com:

    Physics Interactives:

        Circular and Satellite Motion

        Uniform Circular Motion

        Race Track

Roller Coaster Model

Roller Coaster Design

Orbital Motion

Gravitation

The Value of g

The Value of g on the Other Plants

Your Weight on Other Planets

Concept Builders:

Circular and Satellite Motion

Circular Motion

Case Studies-Circular Motion

Forces and Fee-Body diagrams in Circular Motion

Universal Gravitation

Gravitational Field Strength

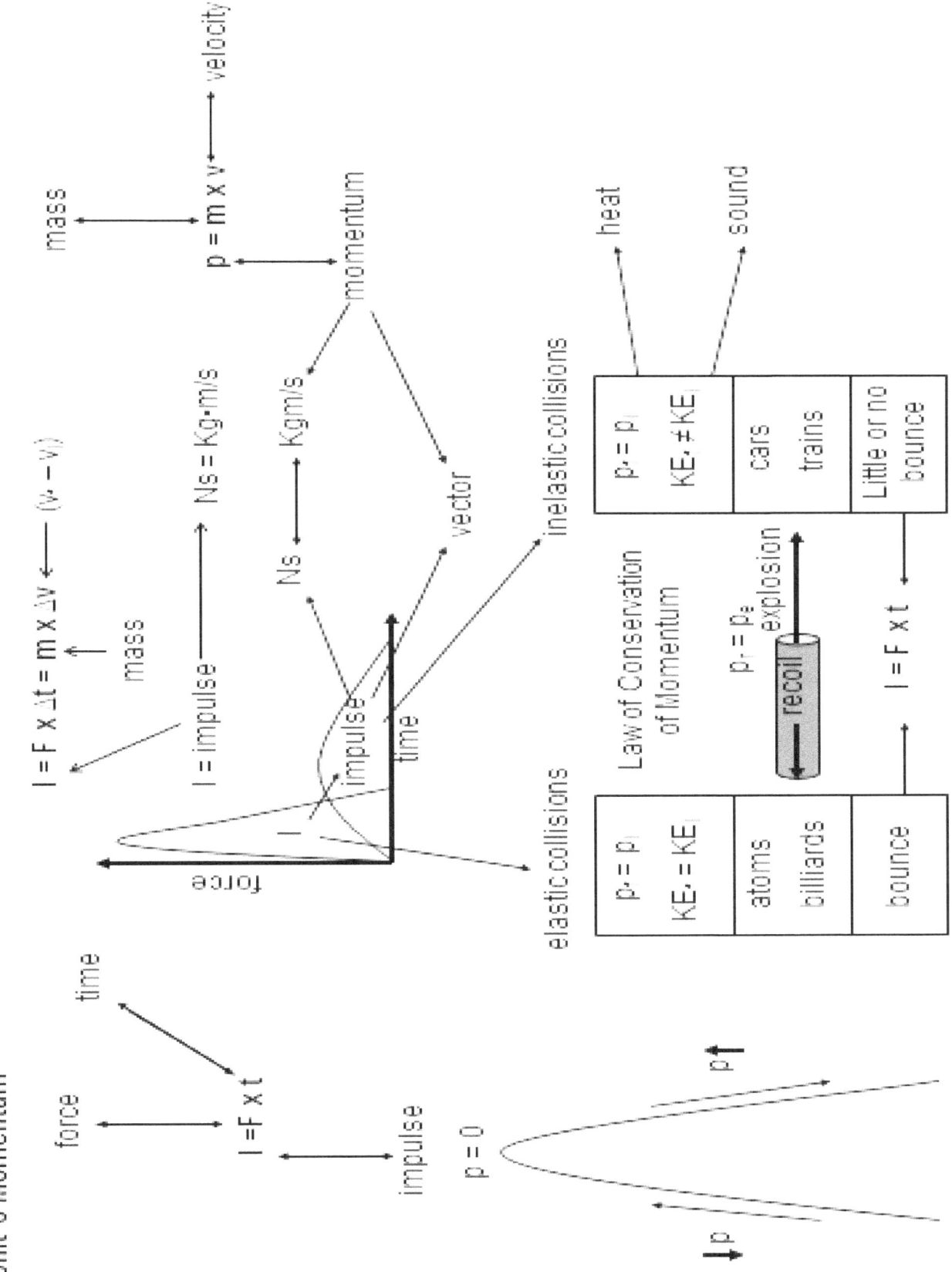

Velocity and Momentum

Directions:

You will need a **scale**, a **stopwatch**, **masking tape**, a **cart**, a **string** (about 1 m long), and a **100 g mass. Looking at the materials and lab we will be using, what are the safety precautions we should take to protect ourselves and materials during the investigation?**

1) Find the cart's mass and write that in the data on the next page.

2) Tie one end of the string to the cart and the other end of the string to the 100 g mass.

3) On your lab table, measure and mark out a starting point for your cart, 20 cm, 40 cm, and 60 cm (right before the edge of the table).

4) Hang the 100 g mass over the table's edge and place and hold the car at the starting point you marked with tape.

5) Let go of the cart and time how long it takes to move from the starting point to the 20 cm mark. Stop the cart and do not let it roll off the table. Place the time data in Data Table 1.

6) Repeat steps #s 4 and 5 for two more trials and write that data in Data Table 1.

7) Repeat the procedure for #s 4-6 for 40 cm and 60 cm distances.

8) Calculate the average time (add each of the three times together and divide by 3) for each distance and put that data in Data Table 2.

9) Calculate the average velocity by taking the distance divided by time.

10) Calculate the final velocity by multiplying the average velocity by 2.

11) Then calculate the momentum by multiplying the cart's mass by its final velocity (g x m/s).

12) Plot a graph for Time vs. Momentum in Graph 1.

Data:

Mass of the cart _____ g

Data Table 1

Distance (meters)	Time trial 1 (seconds)	Time trial 2 (seconds)	Time trial 3 (seconds)
.2 m			
.4 m			
.6 m			

Data Table 2

Distance (meters)	Average Time (seconds)	Average Velocity (m/s)	Final Velocity (m/s)	Momentum (g x m/s)
.2 m				
.4 m				
.6 m				

Graph 1

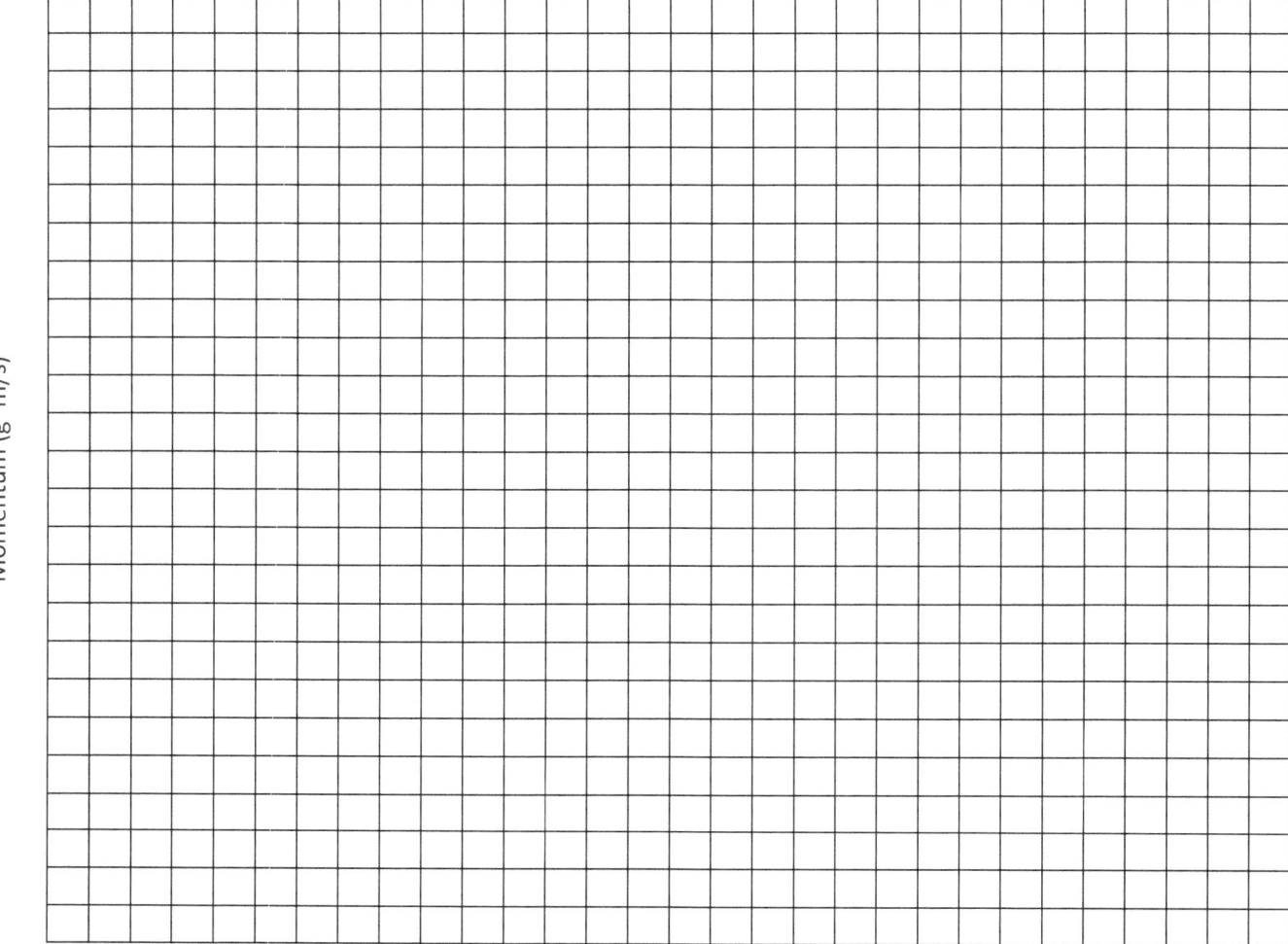

Questions:

1) What happened to the momentum the farther the cart was allowed to move?

2) What kind of movement is happening to the cart?

3) What force caused the cart to move forward?

4) Why do we need to have a constant force acting on the cart?

5) What was the momentum of the cart before it was released?

6) What does the graph indicate about how momentum is related to time when a constant force acts on an object?

7) This information is useful in a golf swing. How could someone easily adjust how far a ball is hit with the same golf club?

8) What do you think is a variable seen here that allows a professional golfer to hit a ball farther than a normal golfer?

The Momentum of Colliding Objects

Directions:

You will need a **scale**, **masking tape**, some type of **ramp** or **track**, a **meter stick**, a **ping pong ball**, a **tennis ball**, a **baseball**, and a **softball. Looking at the materials and lab we will be using, what are the safety precautions we should take to protect ourselves and materials during the investigation?**

1) Find the mass of each object and write them in Data Table 1.

2) Place a piece of tape on the floor and line your ramp/track up to it.

3) Place the softball at the bottom of the track on the tape.

4) Roll the ping pong ball down the track and measure how far it moved the softball.

5) Roll the tennis ball down the ramp and measure how far it moves the softball.

6) Roll the baseball down the ramp and measure how far it moves the softball.

Data Table 1

Ball	Mass (g)	How far it moved the softball (cm)	Rank Momentum
Ping Pong			
Tennis			
Baseball			

Questions:

1) Which ball moved the softball the farthest?

2) This ball had the most momentum, which was passed on to the softball. Rank the momentum of the balls at the bottom of the ramp from highest to lowest in Data Table 1.

3) In this experiment, what caused the biggest change in momentum?

4) How else could we change the momentum?

5) Test it to see if it works.

Conservation of Momentum

Directions:

You will need a **scale**, a **200 g mass**, two carts (**one spring cart** (cart 1) and the other with the different polarity of **magnets on each end** (cart 2), and a **track** from the **Vernier Dynamics kit**. Make sure the bumpers on cart 2 are magnetic to where when you put two together, they stick, and if you turn it around, they will bounce off; this will help you have elastic collisions when they bounce and inelastic when they stick together. **Looking at the materials and lab we will be using, what are the safety precautions we should take to protect ourselves and materials during the investigation?**

1) Make sure the two carts repel each other.
2) Place cart 2 in the middle of the track. Compress the spring in cart 1 so that when you release it, cart 1 will roll into cart 2. Release the cart. Tell me about the motion of the carts before and after the collision?

3) Fix a 200 g mass onto cart 1 and repeat the procedure in #2. What can you tell me about the movement of the carts before and after the collision?

4) Which cart seemed to move faster?

5) Why do you think that is?

6) Switch the 200 g mass to cart 2 and repeat the procedure in #2 so that they bounce off like before. What can you tell me about the movements of the carts before and after the collision?

7) Turn cart 2 around to where they will now stick together when they collide. Repeat the procedure in #2 so the carts collide. What can you tell me about the movements of the cart before and after the collision?

8) Why do you think this happened?

9) Find the mass of both carts and put those in Data Table 2.
10) Make sure there are bumpers on each side of the track. Place a piece of tape in the middle of the track. Place cart 2 so it bounces off cart 1 at the piece of tape.
11) Repeat the procedure in #2 and time how long it takes for cart 1 to collide with cart 2. Repeat this two more times and write the time data in Data Table 1.
12) This time repeat the procedure in #2 and time how long it takes cart 2 to move from the tape to the other bumper for three trials. Write this time data also in Data Table 1.
13) Calculate the average time for each car in Data Table 1 and write it in Data Table 2.
14) Calculate the average velocity for both carts by measuring the distance they traveled in meters and dividing by time. Write this data in Data Table 2.
15) Calculate the momentum of each cart by multiplying their mass by their velocity. Write this also in Data Table 2.

Data Table 1

Trial	Cart 1 Time (s)	Cart 2 Time (s)
1		
2		
3		

Data Table 2

Carts	Ave Time (s)	Average Velocity (m/s)	Mass of carts (g)	Average Momentum
Cart 1				g x m/s
Cart 2				g x m/s

Questions:

1) How does cart one's momentum before the collision compare to cart two's momentum after the collision?

2) How does this explain the results from procedure #s 2, 4, 6, and 7?

3) How do you think cart 1's momentum will compare to both carts when we collide cart one into cart two, having them stick together?

4) Explain how you could test this? Test it and show your data in Data Tables 3 and 4.

Data Table 3

Trials	Cart 1 alone time (s)	Both carts together time (s)
1		
2		
3		

Data Table 4

Carts	Average Time (s)	Average Velocity (m/s)	Mass of Carts (g)	Momentum of Carts (g x m/s)
Cart 1				g x m/s
Both Carts				g x m/s

Egg Drop

Directions:

Use the knowledge you have learned so far about physics and build a device to hold a raw egg and protect it in a collision with the ground when dropped from a high position your teacher decides. Make sure to practice with your own egg(s); if the egg cracks, you must adjust the design of your structure. Draw and explain how your design works to absorb the energy of the collision so the egg does not crack. Make sure it is ready to test in front of your teacher and class on the date your teacher chooses.

Virtual Investigations that go with Momentum

ExploreLearning.com:

    Sled Wars Gizmo

    Roller Coaster Physics Gizmo

    Air Track Gizmo

    2D Collisions Gizmo

    Crumple Zones Gizmo

PhET.colorado.edu:

    Collision Lab

Physicsclassroom.com:

    Physics Interactives:

        Momentum and Collisions

        Egg Drop

        The Cart & Brick

        Fish Catch

        Exploding Cars

        Collision Carts

    Concept Builders:

        Momentum and Collisions

        Momentum

        Being Impulsive About Momentum Change

        Case Studies: Impulse and Force

        Explosions – Law Enforcement

        Hit and Stick Collision – Law Enforcement

Keeping Track of Momentum – Hit and Stick

Keeping Track of Momentum – Hit and Bounce

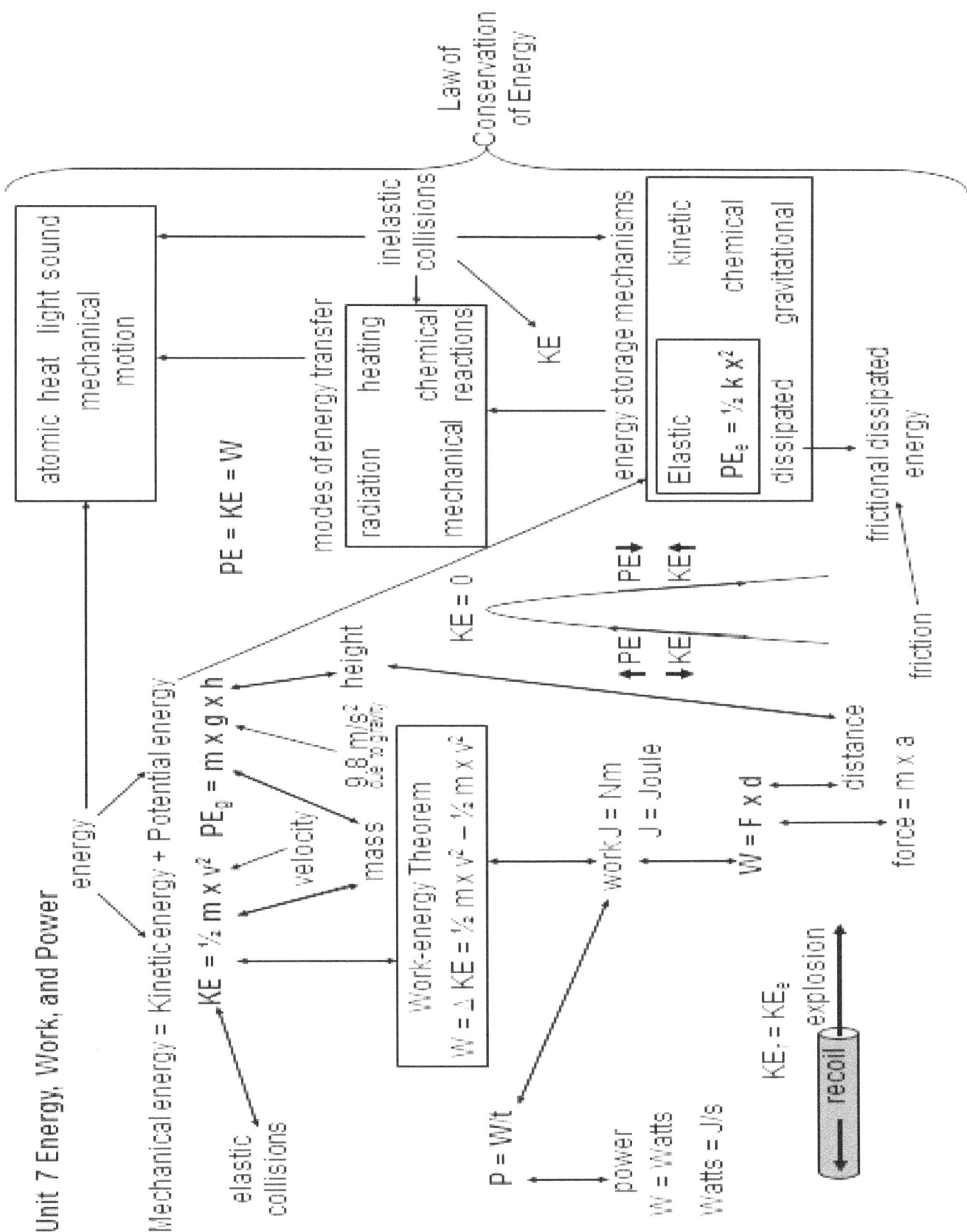

The Energy of a Pendulum Lab

Directions:

You will need a **string**, hanging **masses (10 g, 20 g, and 50 g)**, a **ring stand**, and a **stopwatch**. Looking at the materials and lab we will be using, what are the safety precautions we should take to protect ourselves and materials during the investigation?

1) Take one end of the string and tie it to one side of the ring on a ring stand set as high as you can. Have enough string to where you tie it to the other side of the ring; when you hook the largest hanging mass in the middle of the string, it will be able to swing freely just above the ground/table.
2) Measure the pendulum's length by measuring the distance between the ring's front and the mass's bottom. Write this in the Data Table 1.
3) Have the pendulum swing back and forth and time how long it takes to go through 15 swing cycles with a 50 g mass on it. Write this in Data Table 1.
4) Repeat the procedure in #3 with a 20 g mass on it.
5) Repeat the procedure in #3 with a 10 g mass on it. Write this data in both Data Table 1 and the first part of Data Table 2.
6) Now wind the string around the front of the ring so you shorten the pendulum's length. Measure its length as you did in step #2. Write this measurement in Data Table 2.
7) Repeat the procedure in #3, keeping the 10 g mass on and write that data in Data Table 2.
8) Repeat this procedure in #6-7 one more time for a shorter distance. Write your data in Data Table 2

Data Table 1

Mass (g)	Length of Pendulum (cm)	Time for 15 Cycles
50 g	cm	s
20 g	cm	s
10 g	cm	s

Data Table 2

Mass (g)	Length of Pendulum (cm)	Time for 15 Cycles
10 g	cm	s
10 g	cm	s
10 g	cm	s

Questions:

1) Where in the swing is the pendulum moving the fastest?

2) Where in the swing is the pendulum moving the slowest?

3) Potential Energy is defined as having the ability to do work. Where do we see the pendulum with the highest Potential Energy?

4) Kinetic Energy is the energy of movement. Where does the pendulum have the highest kinetic energy?

5) How does changing the mass affect the time it takes to complete 15 cycles of the pendulum?

6) How does changing the length of the pendulum affect the time it takes to complete 15 cycles?

7) How do the PE and KE change as the pendulum moves from its highest point to its lowest point?

8) How do the PE and KE change as the pendulum rises from its lowest point to its highest point?

9) Write a formula showing the relationship between Potential Energy (PE), Kinetic Energy (KE), and Total Energy (TE).

10) How does this information relate to the energy change when kicking a ball straight up into the air?

Simple Physics Investigations

Potential and Kinetic Energy

Directions:

You will need a **scale**, a **golf ball**, a **tennis ball**, and a **meter stick**. **Looking at the materials and lab we will be using, what are the safety precautions we should take to protect ourselves and materials during the investigation?**

1) Find the mass of the golf ball and tennis ball and convert them from grams to kilograms. Write this in Data Table 1.
2) Measure the height of the lab table in meters. Write this in Data Table 1.
3) Calculate the Potential Energy of the balls using the formula PE = m x g x h, where PE is measured in Joules (J), mass is measured in kilograms (kg), acceleration due to gravity is 9.81 m/s², and height is in meters (m). Write this in Data Table 1.

Data Table 1

Object	Mass (g)	Mass (kg)	Location	Height (m)	PE (J)
Golf Ball			On Floor	0 m	
Golf Ball			On Table		
Tennis Ball			On Floor	0 m	
Tennis Ball			On Table		

4) Push the golf ball off the edge of the table and time how long it takes for the ball to fall to the floor. Repeat this two more times and place the data in Data Table 2.
5) Repeat the procedure in #4 for the tennis ball. Calculate the average times for both the golf and tennis balls by adding the three numbers and dividing by 3.
6) Take the distance and divide by the average time to get the average velocity.
7) Multiply the average velocity by 2 to get the final velocity.
8) Calculate the kinetic energy of the balls when they hit the ground using the formula KE = ½ x (m x v²), where kinetic energy is measured in joules (J), mass is measured in kilograms (kg), and velocity is measured in meters per second (m/s).

Data Table 2

Object	Trial 1 Time (s)	Trial 2 Time (s)	Trial 3 Time (s)	Average Time (s)	Average Velocity	Final Velocity	KE (J)
Golf Ball					m/s	m/s	
Tennis Ball					m/s	m/s	

Questions:

1) How does the PE compare to the KE?

2) Why did we double the average velocity of a dropped object from rest to get the final velocity?

3) What happened to the PE and KE as the balls fell?

4) Write a formula for the conservation of energy using TE, PE, and KE.

5) Sketch a graph that shows how energy transfers as objects drop and fall from where they are dropped to where they hit the ground. Draw a line for potential energy, kinetic energy, and total energy

Analyzing Elastic Potential Energy

Directions:

You will need a **rubber band**, a **meter stick**, a **coin**, or some type of **disc. Looking at the materials and lab we will be using, what are the safety precautions we should take to protect ourselves and materials during the investigation?**

1) Make marks on your table every .5 cm for 4 centimeters.
2) Place the rubber band on the zero mark so that the rubber band has no slack between your fingers.
3) Place your coin or disc in front of that rubber band, pull it back .5 cm, and release it.
4) Measure how far the disc traveled. Put this data in Data Table 1 below.
5) Repeat the procedures in #s 2-4, pulling the disc back 1 cm this time, and place this data in Data Table 1 below. Repeat this at .5 cm intervals, going longer until you pull it back 4 cm.

Data Table 1

Length Pulled Back (cm)	Distance Traveled (cm)
.5 cm	
1 cm	
1.5 cm	
2 cm	
2.5 cm	
3 cm	
3.5 cm	
4 cm	

Questions:

1) How did the disc's takeoff speed seem to change as you pulled the disc back farther and farther?

    a. What does this imply about the kinetic energy of the disc when released?

    b. What does this imply about the potential energy as you pull the rubber band back?

2) How does this investigation show the conservation of energy?

3) Write an equation that shows this?

4) What causes the disc to stop?

5) Where does that energy go as it stops?

6) How does this relate to how we slow down spacecraft that renter the atmosphere?

    a. What precautions do we have to take?

Simple Physics Investigations · Seven Sides Publishing

Happy and Sad Balls

Directions:

You will need a **scale** and a **pair of rubber balls**; one that is **happy (bounces)** and the other that looks just like it that is **sad (does not bounce)**. You will also need a **meter stick** and a **plastic shoebox** with a **lid**. **Looking at the materials and lab we will be using, what are the safety precautions we should take to protect ourselves and materials during the investigation?**

1) Find the mass of both balls: Happy ball _____kg Sad ball _____kg
2) Hold the meter stick vertically on the table so that 0 cm is on the table and 100 cm is in the air.
3) Hold the happy ball one meter off the table next to the meter stick and drop it so that it bounces next to the meter stick. Measure the height of how high it bounces. Write this in Data Table 1.
4) Repeat the procedure in #3 two more times and enter the data in Data Table 1.
5) Now place the box on the table and the meter stick on the box. Hold the happy ball one meter above the box next to the meter stick, drop it so that it bounces off the box, and measure how high the ball bounces. Write this information in Data Table 1.
6) Repeat the procedure in # 5 two more times and write that data in Data Table 1.
7) Now take the sad ball and repeat the procedures in #s 2-3, bouncing it on the table and write that data in Data Table 1.
8) Repeat #7 two more times and write the data in Data Table 1.
9) Calculate the bounce trials' average height in Data Table 1 by adding the three bounces and dividing by 3 for each ball to surface set-up.

Data Table 1

Ball	Surface	Trial 1 Bounce (m)	Trial 2 Bounce (m)	Trial 3 Bounce (m)	Average Height (m)
Happy	Table				
Happy	Box				
Sad	Table				

Questions:

1) Calculate the gravitational potential energy of the happy ball.

2) Calculate the gravitational potential energy of the sad ball.

3) What happens to its gravitational potential energy and kinetic energy as the ball falls?

4) What happens to their kinetic energy when they hit the floor?

5) How did elastic potential energy help it bounce?

6) How did the bounce heights compare when dropped on the box instead of the floor?

7) Why did this happen? Did you hear anything or see anything move?

8) Why do you think the sad ball did not bounce as high?

9) Did the sad ball have much energy transfer into elastic potential energy?

10) How does the information from this lab help us understand how bulletproof vests work?

Conservation of Energy in a Toy

Directions and Questions:
What you are going to do today is called reverse engineering. You will look at **toys using simple machines**, see how they work, and trace how energy is transferred through the toy from when it enters your toy until it leaves your toy. You will describe what simple machines or devices are used to make the toy move for its function. **Looking at the materials and lab we will be using, what are the safety precautions we should take to protect ourselves and materials during the investigation?**

1) Describe in as much detail as you can what your toy does.

2) How does the energy enter your toy?

3) Describe how the energy is transferred through your toy (tell what machines or devices are used).

4) Tell all the different ways energy leaves your toy.

5) Leave your toy with this paper and look at someone else's toy and see how accurate they were, and add observations and corrections as they do the same to yours. Discuss the changes/additions you made to each other's work (collaboration).

Energy and Rockets Lab

Materials and Safety:

You will need a **stomp air rocket launcher**, **2 lb.** and **8 lb. medicine balls**, and measure your **rocket's mass** with a **scale. Looking at the materials and lab we will be using, what are the safety precautions we should take to protect ourselves and materials during the investigation?**

Rocket's mass: _____

Prediction:

1) If you drop the same medicine ball from a high height and a low height, which height should make the rocket fly higher?

Experiment:

2) Drop the same medicine ball from two significantly different heights on the stomping part of the rocket launcher. Which height made the rocket fly the highest?

Conclusion:

3) Why did that height make the rocket fly higher than the other?

Prediction:

4) Which will cause the rocket to fly higher if dropped at the same height, the 2 lb. medicine ball or the 8 lb. medicine ball?

Experiment:

5) Using the same rocket launcher, drop each medicine ball on the rocket launcher from the same height and see which ball caused the rocket to fly higher. Which ball caused the rocket to fly higher?

Conclusion:

6) Why did that medicine ball cause the rocket to fly higher?

7) What type of energy was in the medicine ball before you dropped it?

8) What type of energy was in the medicine ball when it hit the rocket launcher?

9) What type of energy did the rocket get when it took off?

10) What type of energy did the rocket have at its highest point?

11) Describe the energy transformations that took place in the medicine ball, the medicine ball to the rocket, and how it transformed into the rocket.

12) Calculate the KE of each medicine ball when it hits the ground if dropped from 1m.
    Hint: 1 kg = 2.2 lb

13) What was the rocket's velocity when it left the launcher for both medicine balls?

Who's got the Power?

Directions:

You will need a **dowel** or **PVC pipe**, a **500 g mass**, **string** about 1m long, **masking tape**, and a **stopwatch. Looking at the materials and lab we will be using, what are the safety precautions we should take to protect ourselves and materials during the investigation?**

1) Tie and tape one end of the string to the dowel/PVC pipe. Measure 75 cm down and put some masking tape as a marker. Hook the 500 g mass to the other end of the string.

2) Take turns to roll the dowel/PVC pipe to 75 cm, pulling the mass up while timing how long it takes with the stopwatch.

Data Table

Measurement	Student 1	Student 2
Time (s)		
Force (N)		
Distance (m)		

Questions:

1) Calculate the work both students did. (W = F x d)

2) Calculate the power of both students. (P = W/t)

3) Compare the amount of work each of you did. Why is the work the same?

4) Compare the power of both students. Who had the most power?

5) Why would the powers differ?

6) How does speed affect the amounts of work and power (trick question)?

7) How does work relate to energy?

8) How does power relate to energy?

Levers Lab

Directions Part 1:

You will need a **meter stick**, **string**, **hanging masses**, a **fulcrum collar**, and a **support stand** for the fulcrum. Make sure the middle of the fulcrum is lined up on the 50 cm mark. **Looking at the materials and lab we will be using, what are the safety precautions we should take to protect ourselves and materials during the investigation?**

1) Take a 50 g mass and some string and hang it 20 cm from the fulcrum. Then hang a 25 g mass on the other end of the meter stick at a distance that allows it to balance the meter stick on the stand with the fulcrum. Use that distance to fill in Data Table 1 for Trial 1.
2) Now move the 50 g mass 30 cm away from the fulcrum. Now take a 100 g mass and place it on the other side of the meter stick at a distance that balances the meter stick again. Fill in Data Table 1 with the measurement for Trial 2.
3) Now take two different masses and line them up on opposite sides of the meter stick to get those to balance the meter stick two more times. Write this information in Data Table 1.

Data Table 1

Trial	Mass on Left	Distance on Left	Mass on Right	Distance on Right
1	50 g	20 cm	25 g	
2	50 g	30 cm	100 g	
3				
4				

Questions Part 1:

1) What pattern do you see with the mass and distance from the fulcrum?

2) Write a formula that tells you how to balance a meter stick with two different masses.

3) If you convert the masses to weight or force, you can find the torque. How would you convert these masses to force?

4) Calculate the force of weight for each of the masses.

5) Now use the formula Torque = force x distance to calculate the torque on each side of the meter stick for each trail.

6) What do you notice about the unit for torque? Where is it also used?

7) How do you move a force to achieve a mechanical advantage?

Directions Part 2:

1) Now take three different masses and balance two of them on the left and one on the right. Write the masses and distances in Data Table 2.
2) Repeat Part 2 procedure #1 three more times with different masses and distances. Write your data to fill in Data Table 2.
3) How do you think the left side will compare to the right?

4) Write a formula for this equality.

Data Table 2

Trial	1st Mass Left	1st Distance Left	2nd Mass Left	2nd Distance Left	Mass on Right	Distance on Right
1						
2						
3						
4						

Questions Part 2:

1) So does your formula work with the results?

2) If not, write a formula that will work with Data Table 2.

3) Use this formula to calculate and prove the equality mathematically for trial 1.

4) Prove the equality for trial 2.

5) Prove the equality for trial 3.

6) Prove the equality for trial 4.

Simple Physics Investigations   Seven Sides Publishing

Simple Machines Lab

Directions and Questions:

You will need a **shoe**, a **cart with wheels**, a **screwdriver**, a **round doorknob**, a **lever doorknob**, and a **dual-range force sensor** attached to an **interface** connected to a **computer** with **Logger Pro**. **Looking at the materials and lab we will be using, what are the safety precautions we should take to protect ourselves and materials during the investigation?**

1) A wheel and axle is a modified lever that allows you to move a force over a longer distance. The bigger the radius, the bigger the mechanical advantage. (**Work = F x d**)
2) Hook your force sensor up to a shoe and slide it across the table to see the average force applied by the force sensor you are pulling on the shoe. What was the force?

3) Now do the same thing as #2 but put the shoe on the cart so it rolls across the table. Which trial required the least amount of force?

   a. Which force is greater, sliding friction or rolling friction?

4) Which trial did more work?

5) Which trial allowed you to do work easier? That is because of the mechanical advantage, allowing you to do the same amount of work but easier over a bigger distance.

6) Look at the **W = F x d** equation. If the work stays the same, what happens to the force if you move it over a bigger distance?

7) Compare a lever doorknob and a circular doorknob. How are the doorknobs alike?

8) How are the doorknobs different?

9) How do you get a mechanical advantage to open a door with the lever?

10) How do you get a mechanical advantage with the round doorknob to open a door?

11) How can a screwdriver give you a mechanical advantage?

12) Which screwdriver would make it easier to screw in a screw, a thick handle or thin?

13) How would a ramp give you a mechanical advantage? Draw it out and explain it below.

14) Explain how the forces on a wedge under a door hold the door back. Draw a diagram and explain it below.

15) How did all of the simple machines make work easier?

16) How does a close-pin/clip work as a simple machine?

17) Find three other examples of simple machines and tell how they work to achieve a mechanical advantage.

Pulley Lab

Directions and Questions:

You will need a **set of hooked masses**, at **least two pulleys**, some **string**, a **ring** clamped onto a **ring stand**, and a **dual-range force sensor** attached to an **interface** connected to a **computer** with **Logger Pro. Looking at the materials and lab we will be using, what are the safety precautions we should take to protect ourselves and materials during the investigation?**

1) A pulley is a wheel and axle, a modified lever that can change the force's direction or increase the mechanical advantage to do work with less force. They can be a single fixed pulley, movable pulley, multiple fixed pulleys, or a block and tackle with a movable pulley.

2) Set up a single fixed pulley on the ring stand. Run a string through the pulley and connect it to the mass. Attach the dual-range force sensor to the other end of the string and pull down, measuring the force required to lift the mass. What was the mass that you picked up?

3) What is the force of the mass (m x 9.81)?

4) What was the force the sensor showed you needed to lift it?

5) Why do you think there was there a small difference in the forces?

6) As you pull down, what direction does the weight go?

7) Now set up a single movable pulley by tying one end of the string to the ring stand and running the string through the pulley hooked to the same mass, and the other end of the string is fixed to the force sensor. What is the force lifting the mass this time?

8) How does this compare to #4?

9) Now set up a block and tackle with one fixed pulley and one movable pulley like the one shown in **Figure 1** below. Make sure the force sensor is pulling like the arrow on the left.

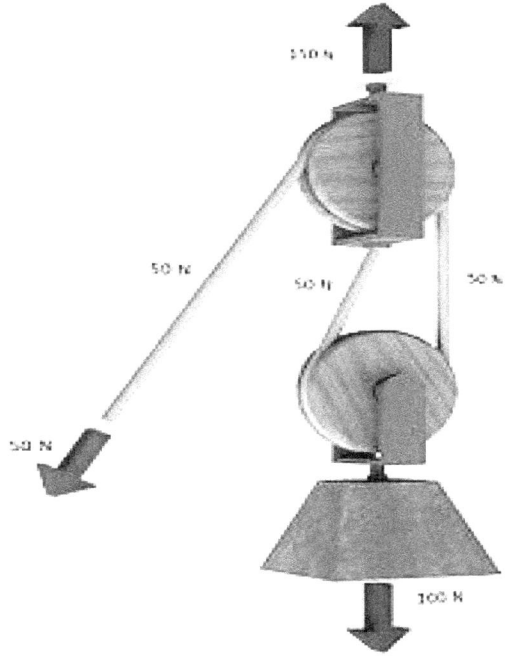

Figure 1: Block and Tackle picture from Creative Commons via Wikipedia.

10) What is the force showing from the sensor?

   a. How does it compare to #s 4 & 8?

11) How do you think we achieved a mechanical advantage in numbers 8 and 10?

12) If you have multiple fixed pulleys, try to set up a system with a higher mechanical advantage. When you do, draw a diagram below showing how you achieved that higher mechanical advantage. Make sure you include what the mechanical advantage was and why.

13) What are some pulleys that we use in our lives for work and recreation?

Simple Physics Investigations   Seven Sides Publishing

Bicycle Lab

Directions:

You will need a **gear-changing bicycle**. Flip the bike upside down, resting on the seat and handlebars. Have the bicycle in the lowest gear. **Looking at the materials and lab we will be using, what are the safety precautions we should take to protect ourselves and materials during the investigation?**

1) Pedal the bike and notice the speed of the back wheel. Now shift it to a higher gear, pedaling the bike at the same pace. How fast is the back wheel moving now compared to when it was in a lower gear?

2) This increase in speed was due to a speed advantage. How do you think that happened?

3) Gears allow us to choose to move vehicles slowly with a high mechanical advantage at a low-speed advantage or fast with a low mechanical advantage but with a high-speed advantage. How do you think this is done while driving the vehicle?

4) Look at the gears around the cranks of the bicycle. Count how many teeth there are around each ring (starting with the biggest gear and ending with the smallest). Write this in Data Table 1.
5) Look at the gears around the back axle of the bicycle. How many teeth are on each gears ring (starting with the smallest and ending with the biggest)? Write them in Data Table 1.
6) Calculate the speed advantage by dividing the front gear by the back gear. Fill in the data table for the combination gears shown in each row.

7) Calculate the mechanical advantage by dividing the back gear by the front gear. Fill in the data table for the combination of gears shown in each row.

Data Table 1

Speed	Front Gear # of Teeth	Back Gear # of Teeth	Speed Advantage	Mechanical Advantage
Fastest	Biggest:	Smallest:		
	Biggest:	2nd Sm:		
	Middle:	3rd Sm:		
	Middle:	3rd Big:		
	Smallest:	2nd Big:		
Slowest	Smallest:	Biggest:		

Questions:

1) How is the mechanical advantage for the fastest speed compared to the mechanical advantage for the slowest speed?

2) How is the speed advantage for the fastest speed compared to the speed advantage for the slowest speed?

3) We only looked at the combination of gears for six speeds. How many speeds does your bike have?

4) Why are these bikes built with so many gears/speeds?

5) When would you use the lowest speed/gears?

6) When would you use the highest speed/gears?

7) Gears are a modified wheel and axle, which is a modified lever. You could even say how the gears are set up on the bike are a modified pulley system. What are some other simple machines that are on this bicycle? Tell what they are and how they function on the bike.

8) How does the speed of a bicycle relate to energy?

9) What kind of force do we need to have high kinetic energy?

Building a Rube Golberg Machine

Directions:

Build a Rube Golberg machine where you use simple machines to cause a chain reaction to do work to complete a simple task of your choice. The longer the chain, the better. Use gravitational potential energy, elastic potential energy, and at least three different types of simple machines (lever, incline plane, wheel and axle, pulley, wedge, gears). Set it up and run it in class or use your phone to record it work properly in one unedited scene at home (it is up to your teacher, which you will do). The more steps you use and the more simple machines you use, the higher the grade. Your teacher will decide the grading scale for this project. Describe how your machine works below.

Virtual Investigations that go with Energy Work and Power

ExploreLearning.com:

    Sled Wars Gizmo

    Roller Coaster Physics Gizmo

    Inclined Plane – Sliding Objects Gizmo

    Energy of a Pendulum Gizmo

    Air Track Gizmo

    Trebuchet Gizmo

    Potential Energy on Shelves Gizmo

    Inclined Plane – Simple Machine Gizmo

PhET.colorado.edu:

    Energy Forms & Changes

    Energy Skate Park

    Hook's Law

    Masses and Springs

    Pendulum Lab

    The Ramp

Physicsclassroom.com:

    Physics Interactives:

        Work and Energy

        It's All Uphill

        Stopping Distance

        Roller Coaster Model

Chart that Motion

Vibrating Mass on a Spring

Concept Building:

Work and Energy

Name That Energy

What's Up (and Down) with KE and PE?

Energy Ranking Tasks

Work

Match That Bar Chart

LOL Charts (a.k.a., Energy Bar Charts)

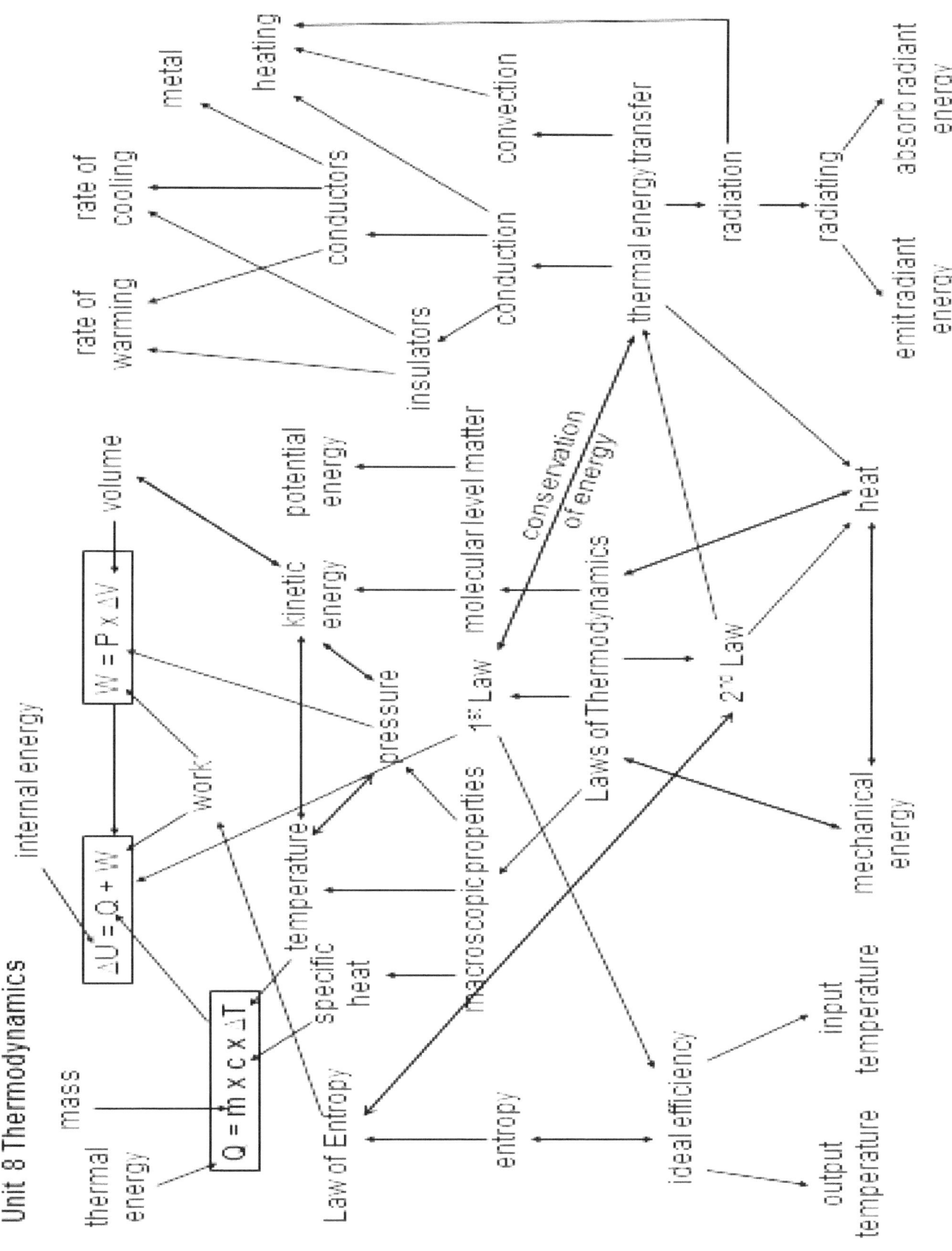

Energy Transformation Balls

Directions:

You will need a pair of **steel energy transformation balls**, an **index card** or piece of **paper**, and **safety goggles. Looking at the materials and lab we will be using, what are the safety precautions we should take to protect ourselves and materials during the investigation?**

1) Put on your safety goggles. Have one person take the steel energy transformation balls and hold one in each hand. Have another person vertically hold out a piece of paper or an index card. The person holding the energy transformation balls should then smash the two balls together on the paper with a very strong force. Make sure not to get anyone's fingers in the way.

2) Observe what happens to the paper where the balls hit and what you smell.

3) Discuss with your teacher the questions that follow.

Questions:

1) What do you see on the paper where the balls hit?

2) What do you smell?

3) What do you think happened?

4) How was energy transformed?

5) How is this collision like a meteor hitting the Earth?

6) How does this relate to the Theory of Relativity E = mc²?

7) How does this relate to the Big Bang Theory?

Student Atomic Motion

Directions and Questions:

You will need a class of **students** (that will act like atoms as a solid, liquid, and gas). **Looking at the materials and lab we will be using, what are the safety precautions we should take to protect ourselves and materials during the investigation?**

1) Have students sit in their seats. They are now modeling how atoms move when they are in a **solid state**. Are they absent of any motion? If not, describe how the student molecules are moving.

2) Now have the students get up and walk slowly around in a small area of the class; they are now modeling **liquid** atoms. How is this motion of the **liquid state** different from the solid state?

3) Now have the students walk around the classroom faster over the entire classroom. When they are about to bump into another person, have them not touch; they just move quickly in another direction. The students are now modeling how atoms move in a **gaseous state**. How do the **gas** atoms move differently from the other two states?

4) Which state of matter had the most energy moving in it? Explain how you can tell.

5) Which state of matter had the least energy moving in it? Explain how you can tell.

6) How could you measure the density of the molecules in this activity?

7) Which state of matter had the smallest density?

8) Which state of matter had the largest density?

9) Which states of matter could change shape to fill the container?

10) Which state of matter could not change shape?

11) Which state of matter was able to fill the whole container?

12) How could we change this model to show molecules (atoms bonded together) in motion as solid, liquid, and gas?

13) How was this model not accurate in showing atomic motion?

Observing Molecular Motion

Directions:

You will need at least three of the **biggest beakers** in your school. They all will be filled with water. One beaker **filled with water** needs to be put in a **refrigerator** so the water is cold. You will also need one beaker on a **hotplate** before doing the demo to heat up (if it boils, you can turn the heat off). The third beaker will have room temperature water fresh out of the tap. Lastly, you will need some **food coloring. Looking at the materials and lab we will be using, what are the safety precautions we should take to protect ourselves and materials during the investigation?**

1) Line all three beakers up from coldest to warmest where the whole class can see. Place a drop of food coloring in the cold beaker and have the students watch how the food coloring spreads.
2) Put drops in the room temperature water, then lastly drops in hot water on the hotplate. Have students observe the movement in all three beakers. It will not take long for the hot one to become homogeneous.
3) Have the students draw what they see in the three beakers below.

Cold   Warm   Hot

Questions:

1) In which beaker did the dye move the fastest?

2) In which beaker did the dye move the slowest?

3) Why do you think the dye moved at different speeds?

4) Temperature is defined as the average kinetic energy of molecules. How does this explain what was happening in the hot, warm, and cold water?

5) Which beaker had the most energy? How do you know?

6) How can this explain why we get hurt when we touch something hot?

Thermal Energy Ventilation

Directions:

You will need a **box**, a **temperature probe** attached to an **interface** connected to a **computer** with **Logger Pro**, and an **incandescent light** (or some type of light bulb that puts out a lot of heat) in a **work light fixture. Looking at the materials and lab we will be using, what are the safety precautions we should take to protect ourselves and materials during the investigation?**

1) Cut two flaps on both sides of the box in the upper half, and cut two flaps on both sides of the box in the lower half of the box. Poke a hole in the top of the box where the temperature probe can snuggly fit.
2) Make sure all flaps are shut, turn on the light inside the box, and close the box. Wait for 5 minutes and measure the temperature inside the box at the end of the 5 minutes. Put that data in Data Table 1.
3) Open the lid and let the heat out. When the temperature goes back to normal, open the bottom flaps on the box, turn on the light, put on the lid, and wait for 5 minutes. When the 5 minutes is up, measure the temperature of the box. Put that in Data Table 1 below.
4) Then repeat #3 except close the bottom flaps and open the top.
5) Then repeat #3 with both sets of flaps open.

Data Table 1

Conditions	Temperature
Both flaps closed	
Bottom flaps open, top closed	
Top flaps open, bottom closed	
All flaps open	

Questions:

1) Which setup had the highest temperature? Why?

2) Which setup had the coolest temperature? Why?

3) Does heat seem to rise or sink? Why?

4) If you had limited space to put vents to release heat, would you put them high on the devise or low? Why?

5) How does this explain why there are vents on many electrical devices?

6) By what method was the inside of the box heated? (conduction, convection, or radiation) Explain.

The Direction Thermal Energy Moves

Directions and Questions:

Place **ice cubes** in a **beaker** of **water. Looking at the materials and lab we will be using, what are the safety precautions we should take to protect ourselves and materials during the investigation?**

1) Thermal energy flows from high energy to low energy. In other words, it moves from warmer areas to colder areas. Write down your observation of what happens to the ice cubes.

2) What was the source of thermal energy?

3) How did the thermal energy move to cause the ice to melt?

4) What gained thermal energy in this system?

5) What lost energy in this system?

6) When it is cold outside, and you open the door to a heated house, in which direction will the energy flow?

   a. Is this accurate when your parents tell you to close the door because you are letting the cold air in? Explain why.

7) When it is warm outside, and you open the door to a cooled house, which direction is the thermal energy flowing?

   a. Is it accurate when your parents tell you to close the door, saying you are letting the cold air out? Explain why.

Convection in Liquids and Gases

Directions and Observations:

Fill a **large beaker** with **water**, add **pepper** to it, place it on a **hotplate,** and heat the water to just below the boiling point. **Looking at the materials and lab we will be using, what are the safety precautions we should take to protect ourselves and materials during the investigation?**

1) Draw a picture of the motion of the pepper in hot water:

2) Describe the motion you see in hot water:

3) What do you think is causing this motion (go into detail)?

4) Light a **candle** with a **match/lighter** and gently blow it out. Which direction does the smoke go?

Questions:

1) Describe how the particles of the pepper moved as the water became hotter.

2) Describe how the pepper particles moved as the water became colder after losing heat on the surface.

3) Explain how convection currents formed in the beaker.

4) Explain why the motion of the particles changed as the burner heated up the water.

5) Which direction did the smoke go when the candle was blown out?

6) Explain why the smoke went in that direction.

Observing Conduction Convection and Radiation

Directions and Questions:

You will need **safety goggles**, a **hotplate**, **Jiffy Pop popcorn**, a **hot air popper**, **unpopped popcorn**, **microwave popcorn,** and a **microwave**. Looking at the materials and lab we will be using, what are the safety precautions we should take to protect ourselves and materials during the investigation?

1) Heat popcorn in the Jiffy Pop skillet on a hotplate. Once the popcorn gets hot, what happens to it?

2) Describe how the energy moves from the hotplate to the popcorn with as much detail as you can.

3) How do you know the energy got there?

4) What method of heat transfer was this?

5) Put popcorn in a running hot air popcorn popper. What happens when the popcorn gets hot?

6) Describe how the energy moved from the popper to the popcorn with as much detail as you can.

7) How did you know the energy got there?

8) What method of heat transfer was this?

9) Microwave a bag of popcorn. What happens to the popcorn?

10) Describe how the energy got to the popcorn with as much detail as you can.

11) How do you know the energy got there?

12) What method of heat transfer was this?

Observing Boyles Law

Directions and Questions:

You will need a **large syringe**, a **small marshmallow**, another **large syringe** but with a **stopcock**, an **industrial suction cup**, a **plastic bottle with a valve fixed to the cap**, an **air pump**, and a **small sealed syringe. Looking at the materials and lab we will be using, what are the safety precautions we should take to protect ourselves and materials during the investigation?**

1) You will need a large syringe and a small marshmallow.
    a. Put a small marshmallow in the syringe. With a large volume in the syringe, put your finger over the opening to block air and squeeze the syringe. What happened to the marshmallow?

    b. What caused this?

    c. Now take your finger off the opening letting the air out. Move the syringe to a low volume, put your finger back over the opening and increase the syringe volume. What happened to the marshmallow?

    d. What caused this?

2) You will need a large syringe and stopcock.
    a. Put the stopcock on the syringe with a large volume in the syringe. Press the syringe making the volume inside smaller. What did you notice about how the pressure felt as you changed the volume?

3) You will need industrial suction cups. Put a suction cup on a very smooth surface. Pull the lever on the cup to increase the air volume inside the cup while it is sealed. Pull up on the cup. What do you notice?

   a. Why do you think this happened?

4) You will need a plastic bottle with a sealed syringe inside that has a valve fixed to the cap to pump air into the bottle from an air pump. Make sure the syringe in the bottle has at least ½ to ¾ of the volume open inside.
   a. Pump air inside the bottle with the syringe inside. What do you notice about the volume of the syringe?

   b. Why do you think that happened?

   c. Now let the air out of the bottle and watch what happens to the volume on the syringe. How did the volume change?

   d. Why do you think this happened?

5) How do volume and pressure affect each other in a sealed system?

Relationship Between Temperature Volume and Pressure: Charles Law and Gay-Lussac's Law

Equipment and Safety:

You will need a **balloon**, **string**, a **meter stick**, a **freezer**, an **Erlenmeyer flask**, a **rubber stopper assembly**, a **gas pressure sensor** and **plastic tubing**, a **temperature probe**, an **interface** connected to a **computer** with **Logger Pro**, two **ring stands** and **clamps**, a **beaker with cold water**, and a **hotplate**. **Looking at the materials and lab we will be using, what are the safety precautions we should take to protect ourselves and materials during the investigation?**

Charles Law: the relationship between temperature and volume in a sealed flexible system.

1) At the beginning of the lab, blow up a balloon, measure its circumference with some string and measure that length with a meter stick. Then put the balloon into a freezer. Come back and check the balloon after the lab.
    a. Circumference before freezer:

    b. Circumference after freezer:

    c. What is the relationship between temperature and volume in a sealed flexible system where pressure does not change? This relationship is Charles Law.

Gay-Lussac's Law: the relationship between temperature and pressure in a sealed system.

2) You will need an Erlenmeyer flask, a rubber stopper assembly that you can seal the flask with, and connect a gas pressure sensor to it with plastic tubing. You will also need a temperature probe. You will connect the gas pressure sensor and temperature probe to an interface, which will be connected to a computer with Logger Pro. You will need a clamp connected to a ring stand to hold the sealed Erlenmeyer flask inside a beaker of cold water on a hotplate. You will need another clamp and ring stand to hold the

temperature probe suspended in the water, not touching any glass during the experiment.

    a. In Logger Pro, open the Chemistry with Vernier folder and file #07 Pressure-Temperature. Make sure everything is set up and secure. Press "Collect."

    b. Turn on the hotplate.

    c. Once you see the water is about to boil (when the temperature is close to 100°C), click stop and turn off the hotplate.

    d. In Logger Pro, look at the temperature vs. pressure graph. What is the relationship between temperature and pressure if the volume cannot change?

    e. Try and adjust the graph to see the temperature down to -273°C. What would the pressure be?

    f. Find out the volume of any gas at -273°C (absolute zero). Look it up on the internet.

    g. -273°C is absolute zero where no energy exists. What implications does this have in light of your findings?

    h. What do you think the temperature might be at the bottom of a black hole?

Testing the Rate of Heat Movement

Directions:

You will use a **drinking glass, styrofoam cup**, and a **tumbler like Rtic or Yeti** to hold **ice water** poured from a **pitcher**. Measure the temperature change they go through with **temperature probes** in each container connected to an **interface** that is connected to a **computer** with **Logger Pro**. Plot a graph with **colored pencils** on Graph 1. **Looking at the materials and lab we will be using, what are the safety precautions we should take to protect ourselves and materials during the investigation?**

1) Fill a drinking glass, a Styrofoam cup, and a tumbler with equal amounts of water from the pitcher of ice water.
2) Set the time to collect the data for 30 minutes. Place a temperature probe in each container and start collecting the temperatures of each.
3) Plot three lines on the graph for the temperature data in Logger Pro in Graph 1.

Graph 1

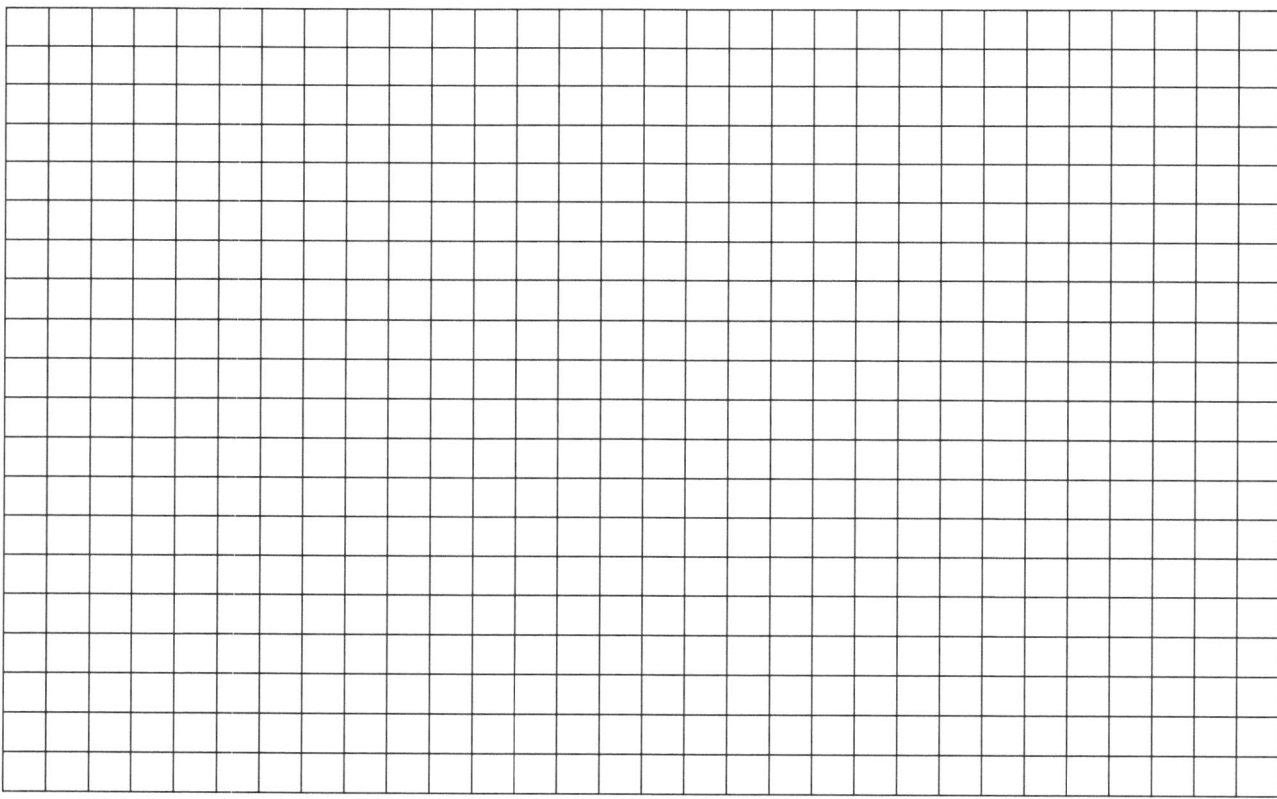

Questions:

1) Why did we use the same amount of water in each container?

    a. Why do you think we did not put ice in the containers?

2) Which container warmed up the fastest?

    a. What do you think about the container that allows heat to transfer in?

3) Which container warmed up the slowest?

    a. What about this container slowed the movement of heat in, trapping thermal energy out of the container?

4) How were conduction, convection, and radiation involved in this investigation?

Virtual Investigations that go with Thermodynamics

ExploreLearning.com:

    Temperature and Particle Motion Gizmo

    Diffusion Gizmo

    Boyle's Law and Charles' Law Gizmo

    Calorimetry Lab Gizmo

    Energy Conversion in a System Gizmo

    Heat Transfer by Conduction Gizmo

    Conduction and Convection Gizmo

    Convection Cells Gizmo

    Radiation Gizmo

    Heat Absorption Gizmo

    Feel the Heat Gizmo

PhET.colorado.edu:

    Balloons and Buoyancy

    Diffusion

    Friction

    Gas Properties

    Gases Intro

    Greenhouse Effect

Physicsclassroom.com:

    Concept Builders:

        Chemistry

        Pressure Concepts

Pressure and Temperature

Volume and Temperature

Pressure and Volume

Unit 9 Electrostatics

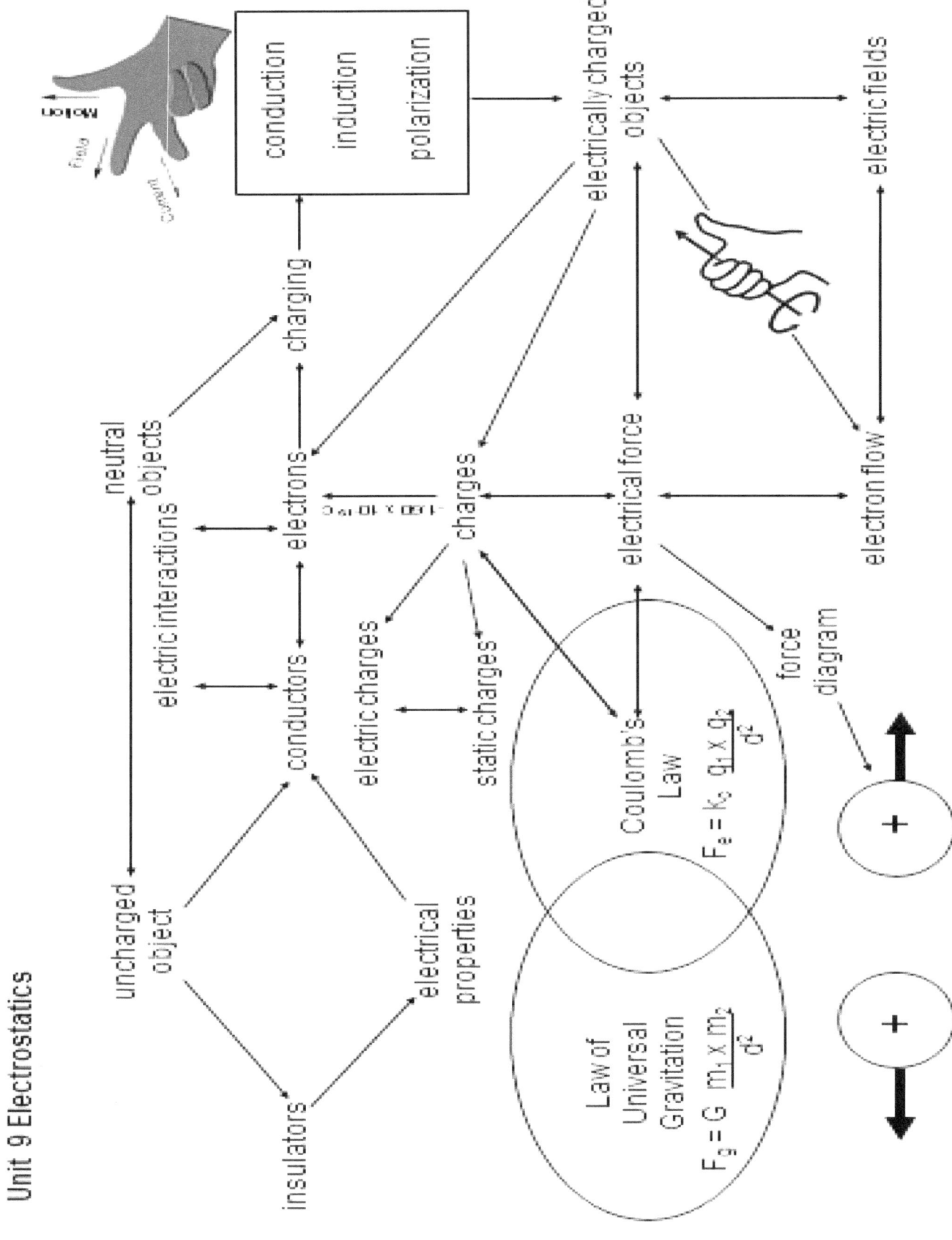

Static Electricity

Directions and Questions:

You will need a **balloon**, an **electroscope**, a **plastic comb,** some **string,** and a **ring stand**. This lab works best when humidity is low. If you do not have an electroscope, one can be made easily with an **Erlenmeyer flask**, a **rubber stopper** with a hole in it, a **screw or bolt** that fits through the hole, a **paper clip** you wind on the threads of the bolt, and two small pieces of **aluminum foil** you hook on the paper clip that will go inside the Erlenmeyer flask. **Looking at the materials and lab we will be using, what are the safety precautions we should take to protect ourselves and materials during the investigation?**

1) Electrons have a negative charge. Since they are the particles that move outside the atom, these are the ones that we can steal off objects. Take your balloon and rub it in your hair; this steals the electrons from your hair and collects them on the balloon. Now move the balloon away from your hair and back towards it. What happens to your hair as the balloon moves towards it?

2) Why do you think this happens?

3) What charge is your hair?

4) What charge is the balloon?

5) Opposite charges attract, like charges repel. See if the balloon will stick to a wall, your shirt, or under a shelf; if they don't, just rub the balloon in your hair some more. Why do you think the balloon sticks?

6) It is easy to move the electrons from the balloon to metal. Take your charged balloon and move it toward the end of the bolt on the electroscope. What happens to the aluminum foil leaves? Why do you think this happens?

7) Pull the balloon away; what happened?

8) Now touch your charged balloon to the bolt on the electroscope; this should transfer the charge. Move the balloon away. What happened to the leaves of aluminum foil?

9) Now touch your finger to the bolt of the electroscope. What did the leaves do now? Why do you think that happened?

10) Now take the plastic comb and tie it to the ring stand. Run the comb through your hair a few times. Then rub the balloon on your head. Bring the balloon to the comb. How does the comb react to the balloon? Why do you think that happens?

11) Now take the charge away from the comb by touching it with your hand. Then bring the charged balloon back to the comb. How did the comb react this time? Why do you think this happened?

12) If you have a faucet, make a small stream of water come out of it. Bring the charged balloon near it. How does the water react to the charged balloon?

13) Water molecules are polar, like magnets. Why do you think the water reacted that way?

14) How does static electricity explain why lightning happens?

15) Did you see any tiny lightning today?

16) How do you think lightning happens?

The Spinning Match

Directions:

You will need two **nickels**, a **match**, a **clear plastic cup**, and an **inflated balloon**. Looking at the materials and lab we will be using, what are the safety precautions we should take to protect ourselves and materials during the investigation?

1) Take the first nickel and place it on a flat surface, tail up. Take the second nickel and balance it on its side on top of the first nickel.
2) Balance a match on top of the two nickels. Place the plastic cup over the set-up.
3) Rub the balloon on your shirt to collect a negative charge on it.
4) Move the balloon near and around the cup.

Questions:

1) What happens to the match?

2) Why do you think this is happening?

3) What do you think is the charge on the tip of the match?

4) What forces are acting on this set-up?

Charged Tape

Directions and Questions:

You will need two **scotch tape** pieces about 10 cm in length with one end folded over to make a handle. **Looking at the materials and lab we will be using, what are the safety precautions we should take to protect ourselves and materials during the investigation?**

1) Stick one strip on your desk/table. Take the other strip and stick it on top of the first. Pull both pieces off together, then pull them apart.
2) Bring the non-sticky sides of both pieces of tape together. What do you see happen?

3) Are their charges alike or different?

4) Now place both pieces of tape directly on the table and peel them off.
5) Bring the non-sticky sides together again. What do you see happen this time?

6) Are the charges alike or different on the tape?

7) What do you think caused the change of the tape?

8) When were the electrons stolen?

Van De Graaff

Directions and Questions:

You will need a **Van De Graaff**, a **golf club iron**, **rubber-soled shoes,** and someone with big hair. **Looking at the materials and lab we will be using, what are the safety precautions we should take to protect ourselves and materials during the investigation?**

1) Turn on the Van De Graaff and bring the golf club near the Van De Graaff (make sure you are holding on to the rubber grip when you do that). What do you notice?

2) Lightning is static discharge moving a negative charge to a positive charge trying to neutralize the charges.

3) If you are brave and do not have a pacemaker or a heart condition, bring your finger near the Van De Graaf. What happened? How did that feel?

4) Turn off the Van De Graaff. Have someone with big hair with no jell or hair spray touch the Van De Graaf. Make sure they are grounded with rubber-soled shoes or standing off the ground on a plastic stool. Then turn on the Van De Graaff and watch their hair. Why do you think this happens?

5) Bring the electroscope near the Van De Graaff, but not too close. What happened to the Aluminum foil leaves?

Calculating Forces on Charged Particles

Directions and Questions:

Use Coulomb's Law to calculate and answer the following questions.

1) Calculate the force between them if a particle with a +3 µC charge is 5 cm from another particle with a -7 µC charge.

   a. Is this an attractive or repulsive force? Explain why.

2) If two particles of equal charges -6 µC are 6 cm apart, calculate the force between them.

   a. Is this an attractive or repulsive force? Explain why.

3) If you have three particles in line 3 cm apart on the same plane, -2 µC on the left, +4 µC in the middle, and -6 µC to the right, calculate the net force acting on the middle charge.

a. Which direction will the middle charge move? Give evidence and explain why.

b. Which direction do you think the charge on the left will move? Give evidence and explain why.

c. Which direction do you think the charge on the right will move? Give evidence and explain why.

Identifying Conductors and Insulators

Directions:

You will need a **battery**, a **battery pack** with wires exposed at the end, a **Christmas light** cut, and insulation stripped off the wires' ends. Put the battery(s) into your battery pack. Attach one end of the exposed wire of the battery pack to one exposed end of the Christmas light by twisting them together. This light will be used to see if the materials are conductors or insulators. Make sure it works by taking the free, exposed ends of the battery pack wire and light bulb wire and making them touch. If the bulb lights, it works.

1) You will also need a variety of materials like a **penny**, a **wooden spoon**, a **metal spoon** or **fork**, a **paper clip**, **paper**, a **comb**, **aluminum foil**, an **aluminum can** (check both the top of the can and the painted label), a **rubber band**, a **pencil**, and the **pencil lead** of a mechanical pencil (really carbon-graphite). **Looking at the materials and lab we will be using, what are the safety precautions we should take to protect ourselves and materials during the investigation?**

2) Test each of the materials you gathered in #1 by taking the exposed free ends of the wires from the battery pack and the Christmas light and touching them both on the material you are testing at the same time on different ends of the material. If the light bulb lights, it is a conductor because electrons can pass through the material. It is an insulator if it does not light because it does not let electrons pass through the material. Fill in Data Table 1 below, listing conductors and insulators.

Data Table 1

Light bulb lights: Conductors	Light bulb does not light: Insulators

Questions:

1) What pattern do you see in the materials that are conductors?

2) What pattern do you see in the materials that are insulators?

3) What prevents the light bulb from lighting in insulators?

4) What other materials could allow the light bulb to light?

5) What other materials might cause the light bulb not to light?

6) Was there anything that lit the light bulb that surprised you? Discuss this with your teacher.

Virtual Investigations that go with Electrostatics

ExploreLearning.com:

    Coulomb Force (Static) Gizmo

    Charge Launcher Gizmo

    Pith Ball Lab Gizmo

    Electrostatic Induction Gizmo

    Polarity and Intermolecular Forces Gizmo

PhET.colorado.edu:

    Balloons and Static Electricity

    Capacitor Lab

    Capacitor Lab Basics

    Charges and Fields

    Conductivity

    Coulomb's Law

    Electric Field Hockey

    Electric Field of Dreams

    John Travoltage

    Radiating Charge

    Static

Physicsclassroom.com:

    Physics Interactives:

        Static Electricity

        Aluminum Can Polarization

        Charging

Name that Charge

Coulomb's Law

Electric Field Lines

Put the Charge in the Goal

Electrostatics Landscapes

Concept Builders:

Static Electricity

Charge and Charging

Charge Interactions

Triboelectric Charging

Polarization

Charging by Induction

Coulomb's Law

Electric Field Intensity

Unit 10 Circuits

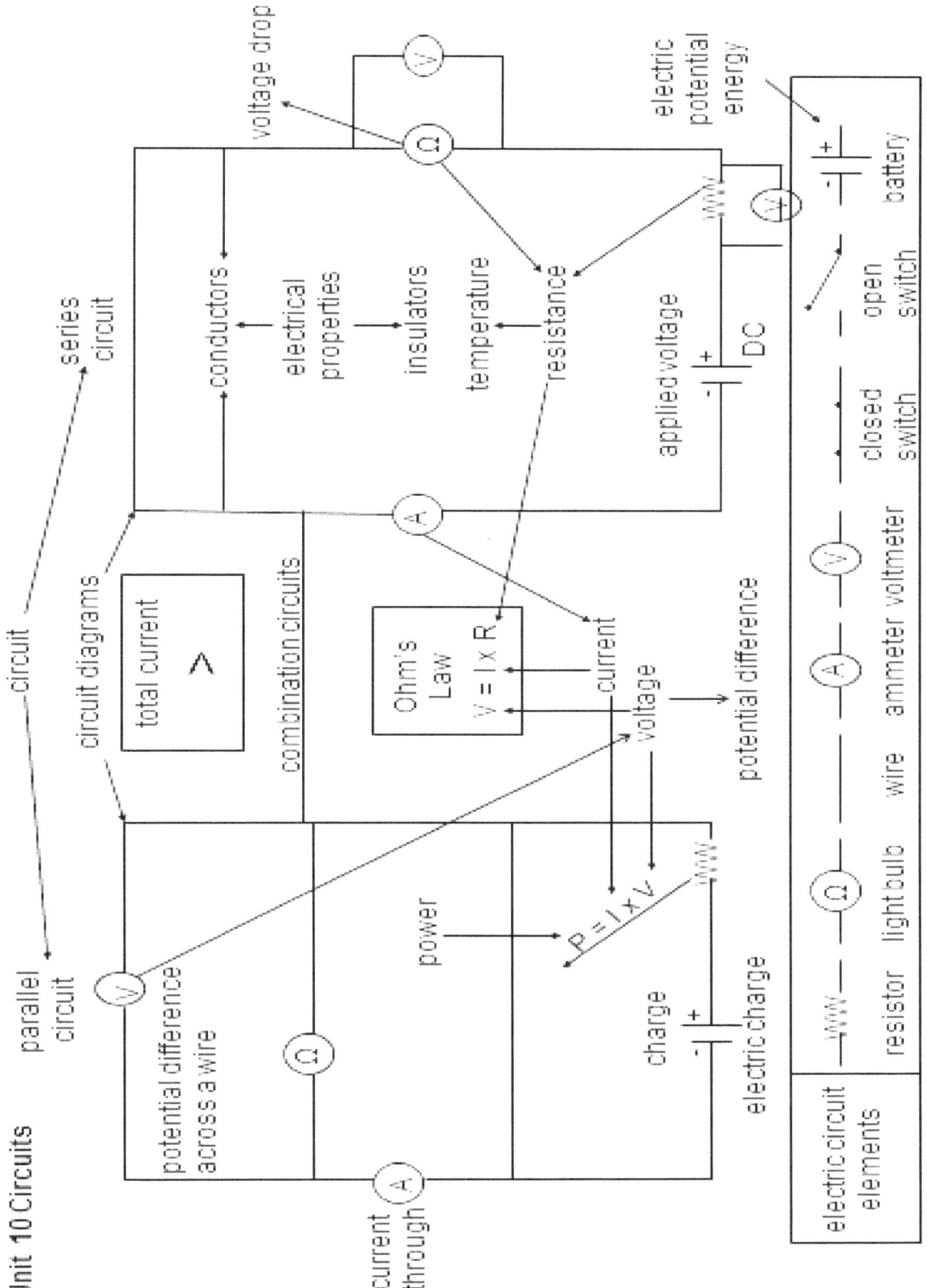

Battery Power

Directions and Questions:

You will need two **batteries** and a string of three **Christmas lights in series** with the ends of the wires exposed. **Looking at the materials and lab we will be using, what are the safety precautions we should take to protect ourselves and materials during the investigation?**

1) Make a circuit lighting the bulbs by putting one wire on the battery's positive end (+) and the other wire on the battery's negative end (−). Notice how bright the lights are.
2) Do the same thing with two batteries stacked together. Which was brighter, one battery or two? Why?

3) What is the voltage difference of each battery?

4) Add them together and find the voltage difference for both together.

5) If a brighter light means a bigger current, what is the relationship between voltage and current?

6) What do you think will happen if you stack three batteries together?

Simple Physics Investigations   Seven Sides Publishing

Making a Graphite Light Bulb

Directions:

You will need **.2 to .5 mm graphite mechanical pencil lead** of different sizes, a **glass jar** with a **lid**, two **wires with alligator clips**, two **6-volt lantern batteries** (total of 12 volts), and **blue tac** (used to fix papers and posters to walls). When you cause resistance in a circuit, heat and light can be given off because the flow of electrons is forced to slow down. **Looking at the materials and lab we will be using, what are the safety precautions we should take to protect ourselves and materials during the investigation?**

1) Take two pieces of the blue tac and fix them to the inside lid of the jar. Place an alligator clip from both wires to the blue tac, holding the mouths up.
2) Take the smallest mechanical pencil lead (graphite) and break a piece off big enough to fit in the two alligator clips and inside the lid of the jar. Place the glass jar over the top of the lid covering the graphite and alligator clips.
3) Turn off the lights in the room. Take the other ends of the wires and clip one wire to the "-" end of the first battery and another wire clipped to the "+" end of the second battery. To complete the circuit, clip one end of another wire on the "+" end of the first battery and clip the other end to the "–" end of the second battery. Watch it light up and notice how bright it is.
4) Make sure to disconnect the clips from the battery setup when you have finished.
5) Repeat the procedures for #s 2-4 for the different size graphite you have. Compare the brightness of each.

Questions:

1) Which size graphite lit up the brightest?

2) Which size graphite lit up the least?

3) Which graphite caused the most resistance to the flow of electrons? How do you know?

4) Which two variables caused the resistance in this investigation?

5) Why do you think the graphite lit up?

6) Look at the periodic table and find Carbon; this makes graphite. Why do you think a nonmetal was able to be used here?

7) Is Carbon a conductor or insulator?

8) What do you see given off when you slow the flow of electrons in a part of the circuit?

9) How do you think an incandescent light bulb lights up when a current flows through it?

10) Explain how electrons and photons are involved with what is happening inside this light bulb.

11) Why did we cover the graphite with the glass jar?

12) What part of Ohm's Law is the graphite in this investigation?

Human Circuits

Directions:

You will need a Current Conductor (found at Hobby Lobby) to show a circuit is closed by lighting up and making noise. **Looking at the materials and lab we will be using, what are the safety precautions we should take to protect ourselves and materials during the investigation?**

Series Circuit:

1) Have students make a circle and hold hands while two of them hold the Current Conductor between them on the metal strips. When everyone is touching hands in the circle, the circuit will be closed, and the Current Conductor will light up and make noise.
2) **Open the Circuit:** Have someone anywhere in the circle let go of someone's hand. What happened to the current of electricity?

3) Have everyone hold hands again, and at a different place, have someone let go. What happened?

4) Do you think this will happen every time in a series circuit?

Parallel Circuit:

5) Have a few students come out of the circle and make one or two lines (depending on how big your class is) connecting across the inside of the circle. Make sure everyone is still touching hands. How do we know the circuit is closed?

6) Have students on the circle opposite to the Current Conductor let go. What happened to the light and sound?

a. Why do you think this happened?

7) Where do people have to let go to make the circuit open? Experiment and find out.

8) If you had appliances in your house hooked up in a series circuit, what would happen if one of them was turned off?

9) How is a parallel circuit different than a series circuit?

10) Do you think your house is wired in series or parallel? Explain why.

Comparing Series and Parallel Circuits

Directions and Questions:

You will need a **multi-meter**, **Christmas lights** in series three with **one bulb**, one with **two bulbs**, one with **three bulbs**, a **battery pack,** and two **batteries. Looking at the materials and lab we will be using, what are the safety precautions we should take to protect ourselves and materials during the investigation?**

Hypotheses:

1) Light bulbs are resisters. They resist or slow the flow of electrons in a circuit, which causes the light bulbs to glow. Predict which lights will glow brighter: three bulbs in series or three bulbs in parallel circuits.

2) Predict which set of lights will stay lit when a bulb is removed: bulbs in series or bulbs in parallel.

Experiment:

3) Compare the lights' brightness when the lights are put into a closed series circuit with 1, 2, and 3 bulbs.

4) Which group of bulbs was the brightest?

5) Which group was the dimmest?

6) Why do you think it happened that way?

7) Compare the brightness of lights when the lights are put into a parallel circuit with 1, 2, and 3 bulbs.

8) Is this the same or different from the series circuit? If different, explain how it is different.

9) Take the three bulbs in series, remove one light carefully, and connect to the battery pack. Do they light up?

10) Take the three bulbs in parallel, remove one light carefully, and connect to the battery pack. Do they light up?

11) Why do you think the results from #s 9 & 10 came out the way they did?

12) How did your results compare to your hypotheses?

Measuring Actual Voltage and Resistance

Directions and Questions:

You will need a **multi-meter**, **Christmas lights** in series three with **one bulb**, one with **two bulbs**, one with **three bulbs**, a **battery pack,** and two **batteries**. **Looking at the materials and lab we will be using, what are the safety precautions we should take to protect ourselves and materials during the investigation?**

Checking Voltage

1) Take a battery out of the battery pack and measure the battery's voltage.

2) Now put the battery back and measure the voltage going through the wires.

3) How is the answer in #2 related to your answer to #1?

4) **Hypothesis.** How do you think voltage will change as you add bulbs in a series circuit?

5) Connect one light to the battery pack and measure the voltage going through one light?

6) Now take off the one light, put on the two lights in series, and measure the voltage going across the lights.

7) Now check the voltage going across three lights in series.

8) **Conclusion.** How are #s 5, 6, and 7 related?

9) **Hypothesis.** How do you think the voltage changes as you add bulbs in a parallel circuit?

10) Check the voltage going across two bulbs in a parallel circuit.

11) Predict what voltage will go across three bulbs in parallel.

12) Now check the voltage going through 3 bulbs in a parallel circuit.

13) **Conclusion.** What do you think you can say about the voltage going through a circuit?

Checking Resistance

14) **Hypothesis.** How do you think the resistance will change as we add bulbs in the series?

15) Check the resistance of one bulb.

16) Check the resistance of 2 bulbs in series.

17) Check the resistance of 3 bulbs in series.

18) **Conclusion.** How does resistance change as you add bulbs in series?

19) **Hypothesis.** How do you think the resistance will change as we add bulbs in parallel?

20) Check the resistance of 2 bulbs in a parallel circuit.

21) Check the resistance of 3 bulbs in a parallel circuit.

22) **Conclusion.** How does the resistance change as you add bulbs in a parallel circuit?

Simple Physics Investigations — Seven Sides Publishing

Hotdog Circuits

Warnings and Directions:

Follow the teacher's directions carefully!

Once you plug in the **extension cord with large alligator clips** attached to each wire, do not touch the **forks** or the **hotdogs**, or you will get shocked. You will also need other **cords with large alligator clips** on both ends but no plug and a **pickle. Looking at the materials and lab we will be using, what are the safety precautions we should take to protect ourselves and materials during the investigation?**

1) Follow the directions to make the systems below, and once plugged in, time how long it takes to cook each hotdog. Write this in Data Table 1. Turn on the stopwatch when you plug in the cord and turn it off when you see the hotdog skin bubble or split. You can then cook longer to your desired taste. **Once your hotdog is done, unplug the cord at the plug before touching anything.**

2) With what you know about circuits, which do you think will cook faster, hotdogs in series or parallel circuits?

3) Have one set-up where you will have one hotdog between two forks with an alligator clip on each fork from the extension cord. Do not plug in the cord until you have set the forks in each end of the hotdogs and the clips are on the forks. **Touching any part of the apparatus (including the hotdog) could cause a damaging shock. Do not touch it until you unplug the set-up. Check with your teacher before you start if you have any questions.**

4) Make another set-up with two hotdogs and another with three hotdogs each in series. You will need to have pieces of cord with no plugs, but alligator clips on each end of the forks put in the hotdogs to connect between the forks of each hotdog so the circuit can run through all the hotdogs. Use the big extension cord with the plug to connect to the hot dogs on each end. **Remember not to plug the cord into the electrical outlet until everything is set up. Once it is plugged in, do not touch it until you unplug each set-up**. Check with your teacher before you start if you have any questions.

5) Make another set-up with two hotdogs in parallel and three in parallel. You do this by poking the forks into each end of each hotdog, then stacking or overlapping the handle ends of the forks over each other, placing them inside each alligator clip of the cord with a plug. **Remember not to plug the cord into the electrical outlet until everything is set up. Once it is plugged in, please do not touch it until you unplug each set-up. Check with your teacher before you start if you have any questions.**

Data Table 1

Number of Hotdogs	Time cooked in series	Time cooked in parallel
1		
2		
3		

Questions:

1) How long did it take to cook one hotdog between two forks?

2) How long did it take for the two hotdogs to cook in a series circuit?

3) How long did it take for the two hotdogs to cook in a parallel circuit?

4) Write down how long it took to cook three hotdogs in series and parallel circuits:

   3 in series: _____ 3 in parallel: _____

5) What part of a circuit does the hotdog represent?

6) What causes the hotdog to cook?

7) If you needed to cook as many hotdogs as you can, as fast as possible, using an electric circuit, which type of circuit would you use and why?

8) Compare and contrast how electricity flows through series and parallel circuits.

Pickle Directions and Questions:

1) Set up a pickle between two forks like you did with the hotdogs. Turn off the lights in the room, and then plug in the cord. Do not touch any part of the apparatus while the pickle is plugged in. What happens when we create a circuit with a **pickle** instead of a hotdog?

2) A pickle is soaked in vinegar (acidic acid); how do you think this affected the electrons in the pickle when plugged in?

3) Do you think what happened with the pickle will happen with a fresh cucumber? Explain why.

4) Explain how the energy changes as electricity is used to light up the pickle?

Virtual Investigations that go with Circuits

ExploreLearning.com:

    Circuit Builder Gizmo

    Circuits Gizmo

    Advanced Circuits Gizmo

PhET.colorado.edu:

    Battery Voltage

    Battery-Resistor Circuit

    Circuit Construction Kit (AC+DC)

    Circuit Construction Kit (AC+DC) Virtual Lab

    Circuit Construction Kit DC

    Circuit Construction Kit DC-Virtual

    Conductivity

    Ohm's Law

    Resistance in a Wire

    Semiconductors

    Single Circuit

Physicsclassroom.com:

    Physics Interactives:

        Electric Circuits

        DC Circuit Builder

    Concept Builders:

        Electric Circuits

        Light Bulb Anatomy

Current

Resistance Ranking Tasks

Know Your Potential

I = ΔV/R Equations as a Guide to Thinking

Which One Doesn't Belong? – Equivalent Resistance

Series Circuits – ΔV = I x R Calculations

Parallel Circuits – ΔV + I x R Calculations

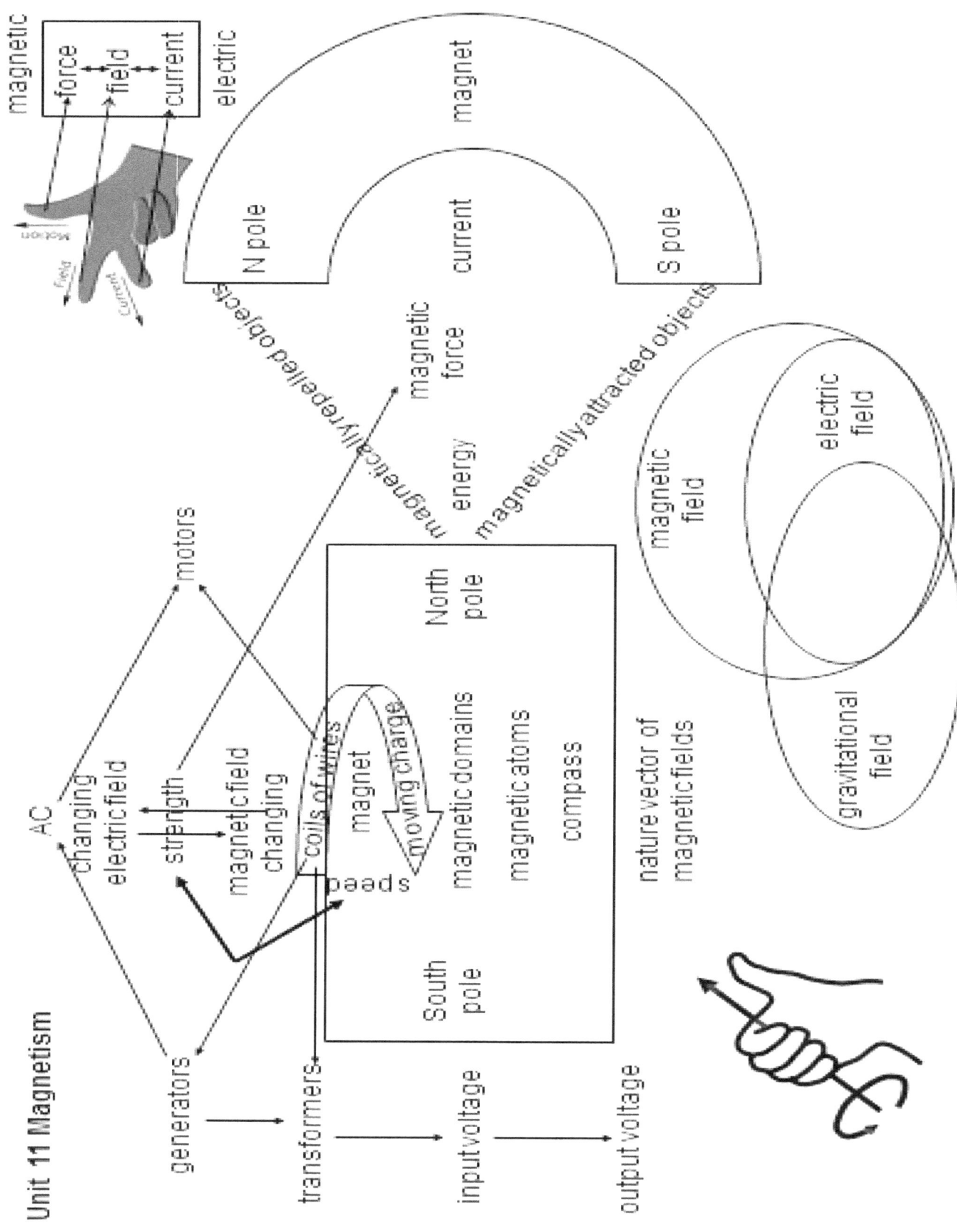

Seeing Magnets

Directions and Observations:

You will need 2 **bar magnets**, a **horseshoe magnet**, and **iron filings** in a **plastic case. Looking at the materials and lab we will be using, what are the safety precautions we should take to protect ourselves and materials during the investigation?**

1) The iron filings align themselves in the magnetic field, allowing us to see the field. Place the bar magnet on top of the plastic case of iron filings and shake them. Draw a picture of the pattern of the iron filings around the magnet.

2) Now place the horseshoe magnet on top of the plastic case of iron filings and shake them. Draw a picture of the pattern of iron filings around the magnet.

3) Now place 2 bar magnets in line so that they are attracted to each other on the plastic case of iron filings and shake them, keeping the two magnets apart. Draw a picture of the pattern of iron filings around the two magnets.

4) Now place 2 bar magnets in line so that they are repelled by each other on the plastic cast of iron filings and shake them, keeping the two magnets apart. Draw a picture of the pattern of iron filings around the two magnets.

Questions:

1) Which seems to have greater strength, the bar magnet or horseshoe?

2) How do you know?

3) How many humps do you see with the attracted magnets?

    a. Is the field connecting or separating between the two magnets?

4) How many humps do you see with the repelling magnets?

    a. Is the field connecting or separating between the two magnets?

5) How does the magnetic field around the attracted magnets appear different than the repelling magnets?

6) Take a bar magnet and tie a string to the middle of it. Hang the magnet away from anything that could attract it. Watch how it orients itself. Place an "N" on the side facing north; this is the magnet's north side. Spin it a little bit and watch it again orient itself. What did you notice?

7) What else is a magnet?

8) Is the North Pole on the Earth magnetically north or south? How do you know?

Making Electromagnets

Directions and Observations:

You will need two **dry-cell batteries**, a **battery pack**, a stronger **wet-cell battery**, two identical **bolts**, thinly insulated **wire**, **BBs** in a **container**, and an **empty container**. **Looking at the materials and lab we will be using, what are the safety precautions we should take to protect ourselves and materials during the investigation?**

1) You will need to take one bolt, wind the wire on all the threads of the screw, and take the insulation off the wire's ends. Dip this into the container of BBs and pull it out. Did any BBs come out?

2) You will need to take the other bolt, wind the wire on half of the threads of the screw, and take the insulation off the wire's ends. Dip this into the container of BBs and pull it out. Did any BBs come out?

3) The bolts with the wire will be the electromagnets. Be careful when you do the next part of the experiment. The setup will get hot fast when you attach the wires to the battery. So you must do the next procedure calmly and quickly, taking a clip off the wire once you are done.

4) Place the half-wrapped bolt into the container with BBs. Attach the battery pack to both ends of the electromagnet wire and gently pull the magnet out of the container with the BBs, trying not to knock any BBs off. Place the electromagnet with the BBs over the empty container and detach one of the electromagnet wires, breaking the circuit.

5) Count how many BBs are in the new container. Give this data to the teacher when called on.

    # of BBs _____ half wrapped electromagnet.

6) Now place the fully wrapped bolt into the container with BBs. Attach the battery pack to both ends of the electromagnet's wires and gently pull the magnet out of the container with the BBs, trying not to knock any BBs off. Place the electromagnet with the BBs over the empty container and detach one of the electromagnet wires, breaking the circuit.

7) Count how many BBs are in the new container. Give this data to the teacher when called on.

    # of BBs _____ fully wrapped electromagnet.

8) Fill in Data Table 1 below as you find out how many BBs each group got for their data in the class.
9) Add the BBs in each column and divide by the number of groups; this will give the average number of BBs for both electromagnets.

Data Table 1

Class Group	# BBs on Half Wrapped Electromagnet	# of BBs on Fully Wrapped Electromagnet
1		
2		
3		
4		
5		
6		
7		
8		
9		
10		
11		
12		
13		
14		
15		
Average # of BBs		

Questions:

1) Did the bolts with the wire have any magnetic properties by themselves? How did you know?

2) Which electromagnet held the most BBs?

3) How is the strength of the magnetic force related to the number of turns on the wire?

4) How does wrapping the wire around the bolt increase the electromagnet's strength?

5) How is the number of BBs related to the magnetic force?

6) Why must the bolts in this experiment be identical?

7) How can electromagnets be more flexible than permanent magnets?

8) What do you think would happen if we use a stronger battery?

9) Have the teacher demo this. How many BBs were picked up this time?

10) How does voltage affect the strength of the electromagnet?

Motors and Generators

Directions and Questions:

You will need various strength **batteries, scrap paper, scissors,** a **simple motor,** and a **flashlight generator.** Looking at the lab and materials we will be using, what are the safety precautions we should take to protect ourselves and materials during the investigation?

Motors

1) Cut out a simple propeller with your scissors that you will push over the motor's axle to spin. Take the smallest voltage battery and touch the wires to the positive and negative terminals of the battery. Which direction did the propeller spin?

2) Switch the wires to touch the opposite terminals of the battery. What direction did the propeller spin now?

3) Notice how fast the propeller spins with this battery. Take the next higher voltage battery and attach the wires to the battery's positive and negative terminals. How fast does the propeller spin now compared to the last battery?

4) How do you think the propeller will spin with an even higher voltage battery?

5) Try it and describe what happened.

6) Describe how you can control a motor's direction and speed. Discuss with your teacher.

Generators

1) Take your flashlight generator and press the handle to make the magnet spin inside the copper wires wrapped around an iron core. What happens to the light bulb as you do this?

2) Speed up how fast you move the magnet in the copper wires. What happens to the light bulb?

3) How do you think electricity is produced? Discuss with your teacher.

4) How else could you move a magnet in copper wires wrapped in an iron core or move copper wires wrapped around an iron core around a magnet?

Comparing Motors and Generators

1) A motor uses a permanent magnet and copper wires around an iron core with a current (which makes an electromagnet) to spin the motor's axle. How is this similar to the generator?

2) Compare and contrast the electric motor and a generator.

3) Could a motor be used as a generator? If so, explain how.

Virtual Investigations that go with Magnetism

ExploreLearning.com:

    Magnetism Gizmo

    Magnetic Induction Gizmo A

    Electromagnetic Induction Gizmo

PhET.colorado.edu:

    Charges and Fields

    Electric Fields of Dreams

    Electric Field Hockey

    Faraday's Electromagnetic Lab

    Faraday's Law

    Generator

    Magnet and Compass

    Magnets and Electromagnets

    Radiating Charge

Physicsclassroom.com:

    Physics Interactives:

        Magnetic Fields

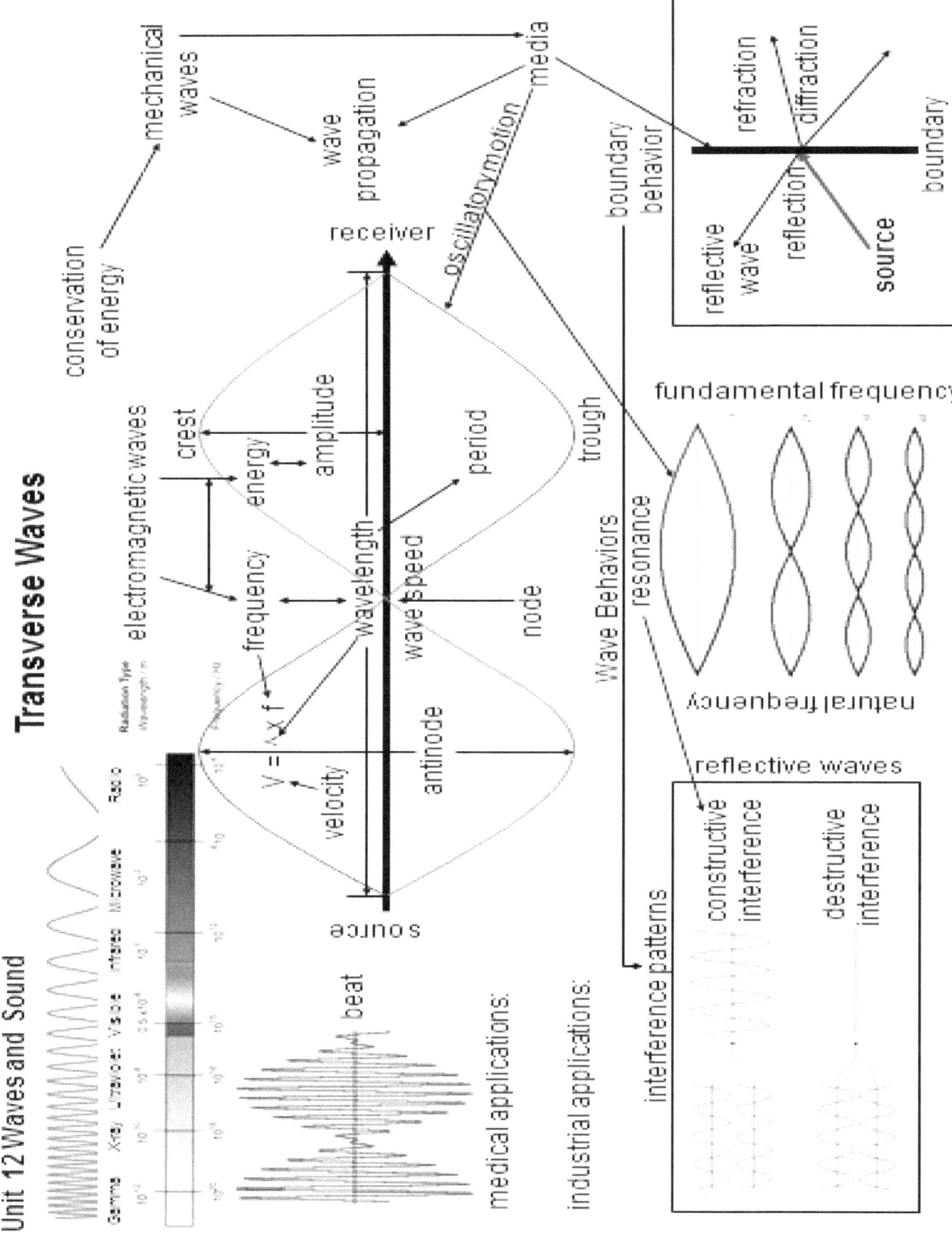

Unit 12 Waves and Sound

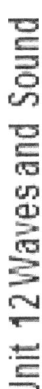

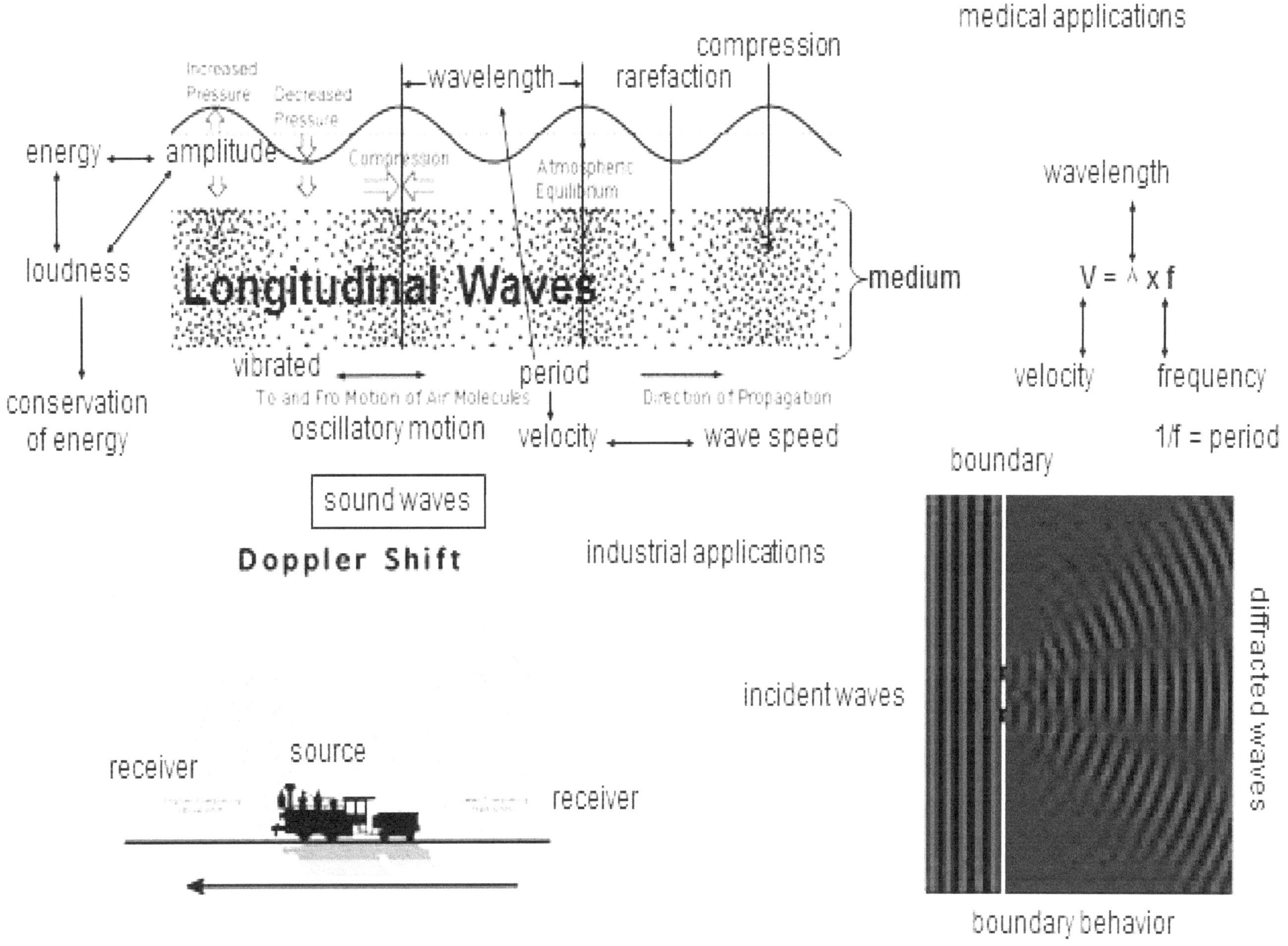

- medical applications
- wavelength ↔ $V = \lambda \times f$ ↔ frequency
- velocity
- $1/f$ = period
- diffracted waves
- boundary
- boundary behavior
- incident waves
- industrial applications
- sound waves
- **Doppler Shift**
- receiver — source — receiver
- compression, rarefaction, wavelength, period, velocity, wave speed
- medium
- Direction of Propagation
- Atmospheric Equilibrium
- To and Fro Motion of Air Molecules
- vibrated, oscillatory motion
- Increased Pressure, Decreased Pressure, Compression
- amplitude
- energy → loudness
- conservation of energy

Longitudinal Waves

Measuring Wave Properties

Directions:

You will need a long springy **telephone cord**, a **stopwatch**, and a **meter stick**. **Looking at the materials and lab we will be using, what are the safety precautions we should take to protect ourselves and materials during the investigation?**

1) Stretch your phone cord and measure how long it is when you will make waves with it.
2) Have two people, one on each end, hold the phone cord. Have one person send a pulse down the cord by having them move their hand back and forth once. Have a third person measure how long it takes for the wave to make it to the other person. In Data Table 1 on the next page, write this data for all three wave times (since all the waves will travel at the same speed down the same medium).
3) Oscillate the cord back and forth to create a standing wave that is half a wavelength long; this will make it look like one hump is in the wave. Make sure you keep the rhythm. In your data table, multiply your cord's length by 2 to get the wavelength for wave one. Write this in Data Table 1.
4) Have the person with the stopwatch; once they see the rhythm on the cord, say start, and they will start the stopwatch and measure 10 seconds and then say stop when the ten seconds are done. The person moving their hand will count how many times they move their hand to the right during the 10 seconds. Write this data down in Data Table 1 on the next page as the wave count.
5) This time oscillate the cord back and forth to create a standing wave that is one wavelength long; this will look like there are two humps. The wavelength for wave 2 is the length of the cord.
6) Repeat the procedure for #4 and write the data for wave 2 in Data Table 1.
7) This time oscillate the cord back and forth to create a standing wave with one and a half wavelengths long; this will look like three humps. The wavelength for wave 3 is 2/3 the length of the cord.
8) Repeat the procedure for #4 and write the data for wave 3 in Data Table 1.
9) Calculate the wave speed by taking the length of the cord and dividing it by the wave time. Write this in Data Table 1 on the next page for all three waves.
10) Calculate the frequency by taking the wave count and dividing it by 10 seconds. Write this in Data Table 1 on the next page for all three waves.

Data Table 1

Wave	Cord Length (m)	Wave Time (s)	Wave Speed (m/s)	Wave Count	Wavelength (m)	Frequency (Hz)
1						
2						
3						

Questions:

1) Why was the wave speed the same for all three waves?

2) What is the relationship between wavelength and frequency?

3) The cord has a natural frequency and what you created were three octaves of that frequency. Where is the word octaves used that has to do with waves?

4) How does frequency relate to pitch?

5) So if someone sings at a high pitch, is the wavelength long or short? Explain why.

6) If someone sings at a low pitch, is the wavelength long or short? Explain why.

Observing Waves in a Slinky

Directions:

You will need a standard to long **slinky** to make both compression waves and transverse waves. You will do this with one person holding one end of the slinky and another person holding the other. One person will move one end; the other will hold still. Each person needs to take a turn moving the slinky to make the waves described below. You will show this to your teacher. **Looking at the material and lab we will be using, what are the safety precautions we should take to protect ourselves and materials during the investigation?**

1) Make a compression wave with high frequency by moving the end of your slinky forward and backward in the same plane as the slinky quickly.

2) Make a compression wave with low frequency by moving the end of the slinky forward and backward in the same plane as the slinky but slowly.

3) Make a compression wave with high amplitude by repeating the procedure in #2 but making bigger, more violent pushes down the slinky.

4) Make a compression wave with low amplitude, the same as the procedure in #2 but making smaller, less violent pushes down the slinky.

5) Make a transverse wave with high frequency by quickly moving the slinky's end perpendicular to the slinky.

6) Make a low-frequency transverse wave by repeating the procedure in #5 but move the slinky more slowly.

7) Make a transverse wave with a high amplitude by repeating the procedure in #5 but moving the end of the slinky a bigger distance perpendicular to the slinky.

8) Make a transverse wave with low amplitude by repeating the procedure in #5 but not as big a distance as #7.

9) Now make a compression wave with high frequency and low amplitude.

10) Now make a compression wave with low frequency and high amplitude.

11) Now make a transverse wave with high frequency and high amplitude.

12) Now make a transverse wave with low frequency and low amplitude.

Observing Sound

Directions:

You will need a **wire hanger** and **string**. Tie two pieces of string on either side of the wire hanger. **Looking at the materials and lab we will be using, what are the safety precautions we should take to protect ourselves and materials during the investigation?**

1) Wrap the string around each of your index fingers and clank the hanger against your desk. How does the hanger sound?

2) Put your fingers in your ears and clank the wire hanger against the desk. How does the hanger sound now?

Questions:

1) Do sounds travel better through the air or the string?

2) Which medium do you think sound travels the fastest through? Put the solid, liquid, and gas in order from fastest to slowest.

3) What if there is no medium for sound to travel through, will there be any sound?

4) When the Death Star blows up in Star Wars, can that really make a sound in outer space? Why?

Coffee Can Phones

Directions and Questions:

Take two **coffee cans** and poke a small hole in the bottom of each of them. Cut a long piece of **string** that reaches across the room, put the ends through each can, and tie a knot in them, fixing them to both cans. **Looking at the materials and lab we will be using, what are the safety precautions we should take to protect ourselves and materials during the investigation?**

1) Pull the string tight while holding the cans and talk through them. Can you be heard in the other can? Why do you think that is?

2) Have someone pinch the string with their fingers halfway across. Can you be heard in the other can now? Why do you think that is?

3) Let the string loosen and droop. Talk again. Can you be heard in the other can? Why do you think that is?

4) Combine with a group next to you, cross your strings, and have them touch while the strings are taught. Have someone talk into a can. Who can hear in their cans?

5) Have someone pinch with their fingers where the strings cross. Can anyone hear now?

6) Explain how sound travels from one can to the other.

Music Test Tubes

Directions:

You will need four **test tubes**, a **test tube holder**, and different **water** amounts placed in each test tube. Test tube one, leave empty. Test tube 2 fill it ¼ full of water. Test tube 3 fill 1/3 with water. Test tube 4 fill with ½ full with water. **Looking at the materials and lab we will be using, what are the safety precautions we should take to protect ourselves and materials during the investigation?**

1) Predict how the toned will sound different when you blow across the top of the test tube. Rank the predicted sounds from 4 being the lowest to 1 being the highest. Write that in Data Table 1 below.
2) Blow across test tube one until you get a tone produced. Do the same for each of the test tubes. Rank the order from the 4 being the lowest tone to 1 being the highest.
3) The tone is created by how large/long the tube is for the air to vibrate. The longer the tube, the longer the wavelength. Empty and wash the test tubes when you are done.

Data Table 1

Test Tube	Amount of Water	Predicted Tone Differences	Tone Difference
1	empty		
2	¼ full		
3	1/3 full		
4	½ full		

Questions:

1) Describe how the tones changed depending on the amount of water in the test tube.

2) How did the pitch depend on the height of the water?

3) Why are the tones different from the different test tubes?

4) Explain how resonance amplifies the sound of a test tube.

5) How do the natural frequencies of the columns of air in each test tube differ?

6) Compare how the test tubes make music with how a flute makes music.

7) How is the flute different from the test tubes?

Singing Glasses and the Dancing Toothpick

Directions and Questions Part 1:

You will need a **flat toothpick**, **crystal glasses of different sizes**, some **crystal glasses of the same size** filled with different **water** amounts, and the **internet. Looking at the materials we will be using, what are the safety precautions we should take to protect ourselves and materials during the investigation?**

1) Wet your finger and rub it around the rim of one of the glasses until you hear a hum.
2) Repeat the procedure in #1 for different sizes of glasses. Which glasses (larger or smaller) have a deeper tone or pitch?

3) Which glasses have a higher tone or pitch?

4) Fill some of the glasses of the same size with different amounts of water. Which glass has the lowest pitch?

5) Which glass has the highest pitch?

6) Research on the internet why your results came out the way they did. What causes the sounds to change when you change the glass's size or how much water is in the glass?

Directions and Questions Part 2:

1) Now take two of the glasses you used in Part 1 that are the same size and place them near each other. Balance a flat toothpick on the rim of one of the glasses.
2) Rub the rim of the other glass to make it hum. What do you notice happens to the toothpick?

3) Discuss with your class and teacher, then explain why this happens.

4) What do natural frequency and resonance have to do with this phenomenon?

Playing the Rubber Band

Directions:

You will need a **plastic tub** and three different **rubber bands** of the same length with different widths. **Looking at the materials and lab we will be using, what are the safety precautions we should take to protect ourselves and materials during the investigation?**

1) Find the mass of each of the rubber bands. Write that in Data Table 1 below.
2) Put the rubber bands around the tub. Try to have the same tension on each rubber band.
3) Pluck each rubber band and tell how they differ. Rank the pitch from 3 being the lowest to 1 being the highest. Put that in Data Table 1 below.

Data Table 1

Rubber Band	Mass	Pitch
Skinny		
Medium		
Thick		

Questions:

1) How does the width of the rubber band affect the pitch?

2) The rubber band that takes the longest to move back and forth will have the lowest frequency. Try to explain why this happens. (Remember the Law of Inertia.)

3) How do you think that affects the acceleration of the rubber bands as they move back and forth?

4) How do you think the length of the rubber band would affect the pitch?

5) How do string thickness and length affect how string instruments sound and are played?

Music has Patterns

Part 1 Directions:

You will need a **digital keyboard** and a **microphone probe** attached to an **interface** connected to a **computer** with **Logger Pro. Looking at the materials and lab we will be using, what are the safety precautions we should take to protect ourselves and materials during the investigation?**

1) In Logger Pro, open the Physics with Vernier folder and file # 35, Mathematics of Music.
2) Position the microphone near the opening where the sound comes out of the instrument. Press "Collect." Play a middle C for the first note. Play it until you see the wave form on the screen. Record the frequency in Hz. Write this in Data Table 1.
3) Now play the next higher note, repeating the procedure in # 2. Repeat this until you have recorded all the notes' frequency going up in the scale until you reach the next C.
4) Now calculate the ratio of the first note to middle C; this is done by taking the current note's frequency and dividing it by the middle C frequency.
5) Take the decimal from the ratio and try to find the fraction that is closest to that decimal, and write your answer like the example shown in Data Table 1. Just like we use different sizes of wrenches in a toolbox or on a ruler in proportions to an inch, you will see a similar pattern. Any variation off of the pattern shows the instrument is out of tune.

Data Table 1

Key	Note	Frequency (Hz)	Ratio to C	Ratio to C Fraction
1	C			1 and 0
2	D			
3	E			1 & 1/4
4	F			
5	G			

6	A			1 & 2/3
7	B			
8	C			

Questions Part 1:

1) What is the frequency ratio to middle C with the next C higher?

2) How long is the wavelength with this C compared to the middle C? Hint: use the wave formula.

3) Are there any other notes that we did not play? What do you think they are?

Part 2 Directions:

1) To see why they are there, play all the keys in order from middle C (even the black keys). Write the frequencies in Hz down in Data Table 2.
2) Then find the ratio to the previous note by taking the current note and dividing it by the frequency of the previous one. What number did you get for each?

Data Table 2

Key	Note	Frequency	Ratio to the prev. note
1 White	C		
2 Black	C sharp		
3 White	D		
4 Black	E flat		
5 White	E		
6 White	F		

7 Black	F sharp		
8 White	G		
9 Black	A flat		
10 White	A		
11 Black	B flat		
12 White	B		
13 White	C		

Questions Part 2:

1) What number did you get for each ratio to the previous note?

2) Besides the pattern of frequency ratios, what other patterns do we see in music?

3) If there are no patterns in the sound, what do we call it?

4) What should you have in sound to be able to call it music?

Observing the Speed of Sound

Directions:

You will need a **thermometer**, a long **tube** about 1 meter long that is open at one end and closed at the other, and a **microphone probe** attached to an **interface** connected to a **computer** with **Logger Pro**. Looking at the materials and lab we will be using, what are the safety precautions we should take to protect ourselves and materials during the investigation?

1) Measure the distance from the opening to the back of the tube; put this in Data Table 1.

2) Find the Temperature of the room in °C and write that in Data Table 1.

3) In Logger Pro, open the folder Physics with Vernier and file # 33 Speed of Sound.

4) Place the microphone directly in the opening of the tube. Click "Collect," and then snap your fingers in the opening, and the data will be taken. You will want to find the time between the two big peaks in the graph; make sure you highlight the same part of each peak. The first peak was the sound and the second peak was the echo; this will be the time the sound took to move from the opening to the closed end of the tube and back to the opening. Write this in Data Table 2 for trial 1.

5) Repeat the procedure for # 4 four more times and write that data in Data Table 2.

6) Average the five trials by adding the times together and then dividing by 5. Write this data in Data Table 2.

7) Multiply the length of the tube by two and use this as the distance the sound traveled.

8) Take the distance from # 7 and divide it by the average time. Write this in Data Table 3 for average speed.

9) Normally the speed of sound is 331.5 m/s at 0°C at atmospheric pressure. The speed of sound increases by .607 m/s for every °C the temperature rises. Find the actual speed of sound in the room and write it in Data Table 3.

Data Table 1

Length of tube	
	m
Temperature of the room	
	°C

Data Table 2

Trial	Travel Time (s)
1	
2	
3	
4	
5	
Average Time (s)	

Data Table 3

Actual Speed (m/s)	Average Speed (m/s)

Questions:

1) Calculate the % accuracy by taking the smaller velocity from Data Table 3, dividing by the larger, and multiplying by 100.

2) Which do you think is faster, the speed of sound or the speed of light?

3) Why is it that when you see a plane flying in the air, the sound seems to be coming from behind the plane?

The Doppler Effect

Directions and Questions:

You will need a toy **football that whistles** when you throw it. Stand between two people that will throw it back and forth over your head. **Looking at the materials and lab we will be using, what are the safety precautions we should take to protect ourselves and materials during the investigation?**

1) When the ball is flying over your head, pay attention to the whistle's pitch. How was the sound before the ball reached you compared to the sound after it passed you?

2) What happens to the pitch of the whistle as it passes over your head?

3) Why do you think this happened?

4) When would the ball move into the waves, shortening the wavelength coming to your ear as it flies in the air?

5) When would the ball move away from the waves, lengthening the wavelength coming into your ear as it flies in the air?

6) How does this explain why a siren sounds higher as it approaches and changes to a lower pitch as it passes?

Virtual Investigations that go with Waves and Sound

ExploreLearning.com:

    Waves Gizmo

    Ripple Tank Gizmo

    Phases Array Gizmo

    Sound Beats and Sine Waves Gizmo

    Longitudinal Waves Gizmo

    Doppler Shift Gizmo

    Doppler Shift Advanced Gizmo

    Earthquakes 1 Recording Station Gizmo

    Earthquakes 2 Determination of Epicenter Gizmo

    Hearing Frequency and Volume Gizmo

PhET.colorado.edu:

    Fourier: Making Waves

    Normal Modes

    Sound

    Wave Interference

    Wave on a String

    Waves Intro

Physicsclassroom.com:

    Physics Interactives:

        Waves and Sound

        Vibrating Mass on a String

        Slinky Lab

Simple Wave Simulator

Wave Addition

Standing Wave Maker

Beat Patterns

Concept Builders:

Waves and Sound

Wave Basics

Wavelength

Waves – Case Studies

Rocking the Boat

Wave Interference

Decibel Scale

Name That Harmonic: Strings

Name That Harmonic: Open-End Air Columns

Name That Harmonic: Closed-End Air Columns

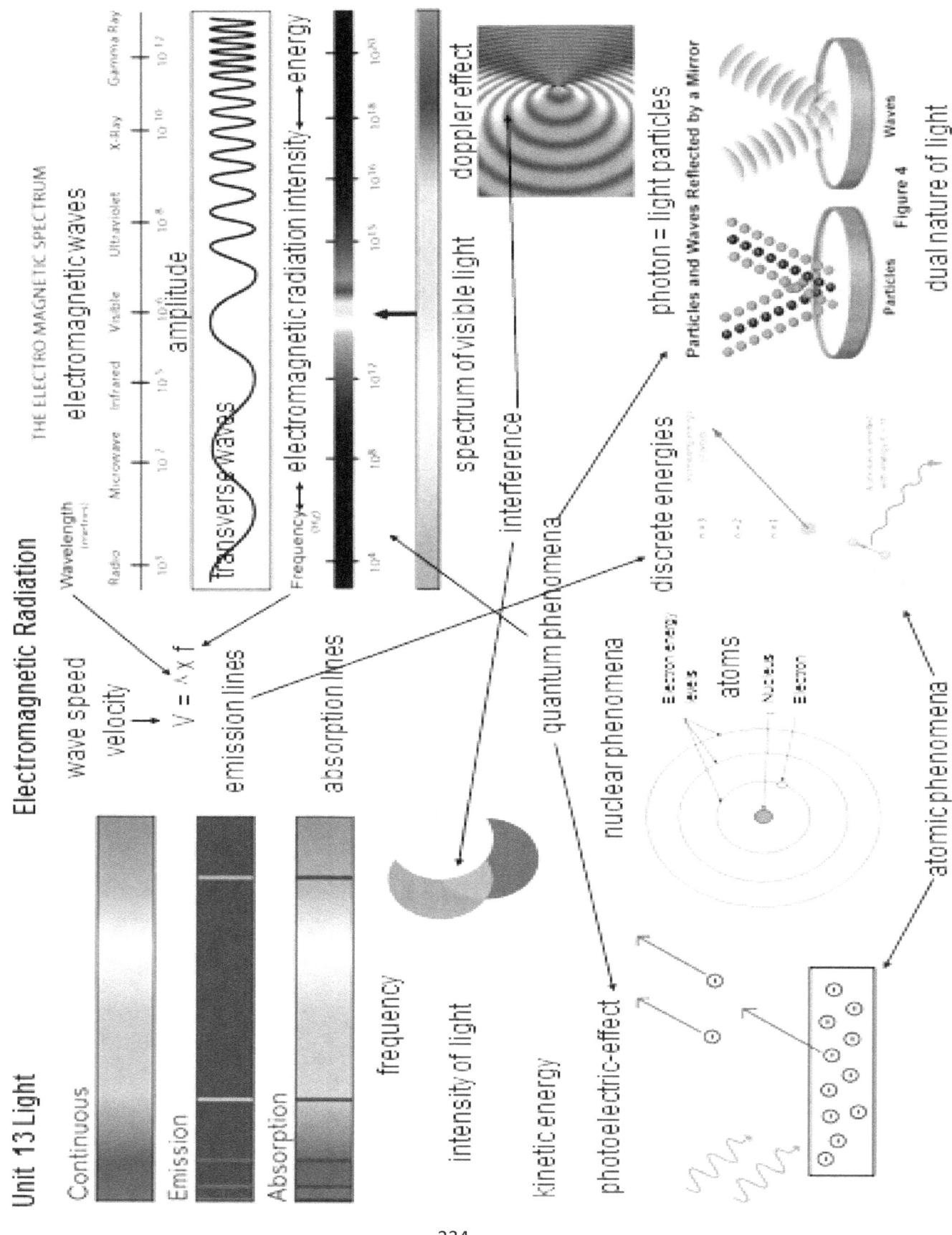

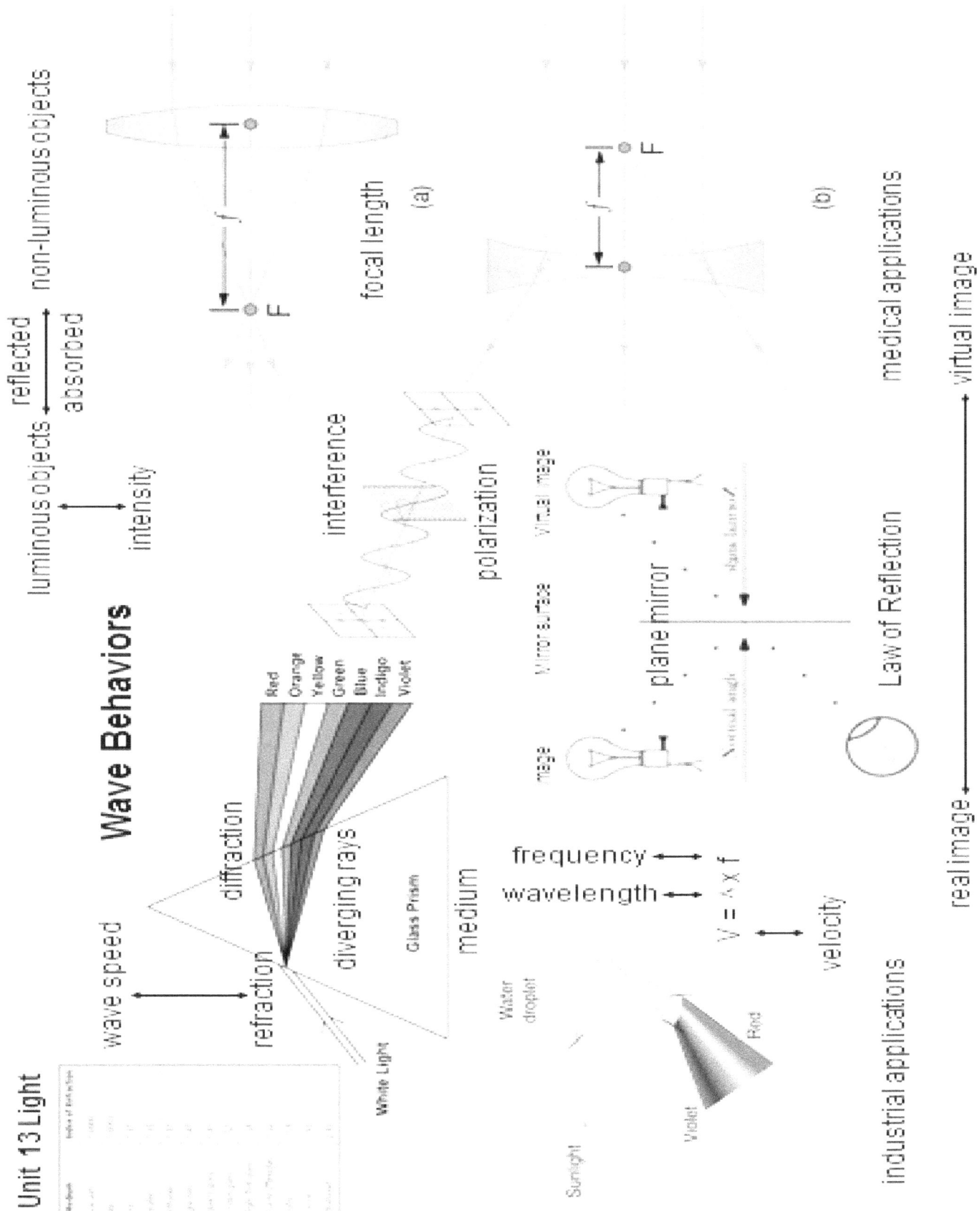

Uses of the Electromagnetic Spectrum

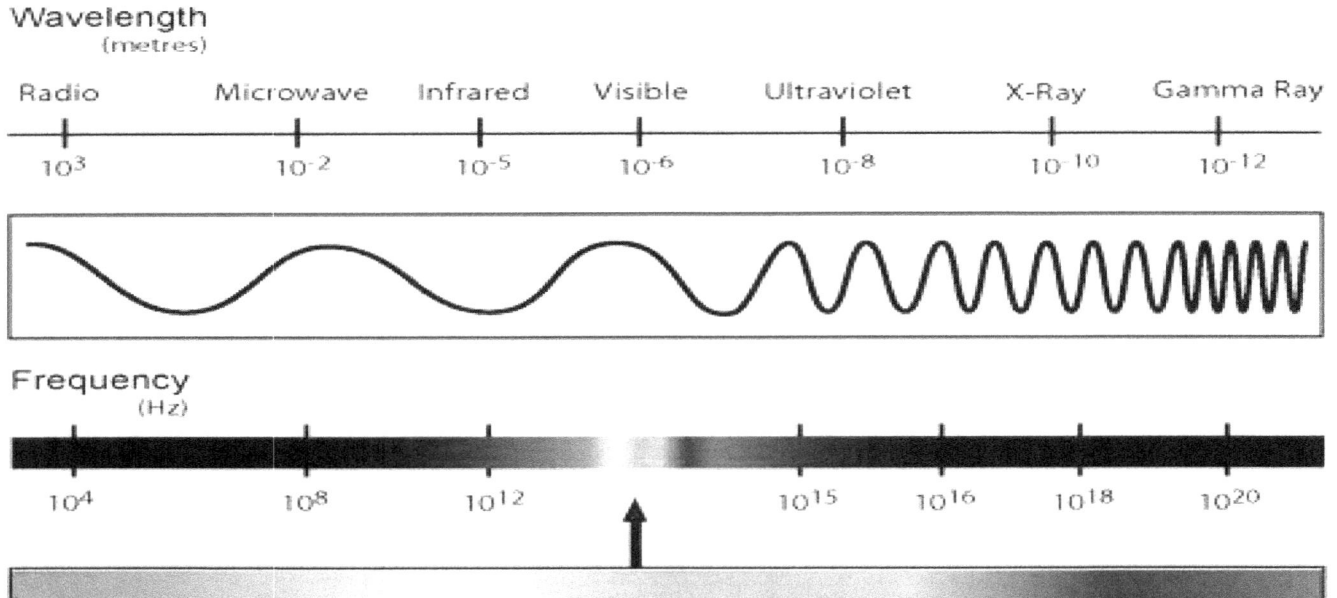

Directions and Questions:

Use the Electromagnetic Spectrum above and the **internet** to answer the questions below.

1) Electromagnetic waves have higher energy with shorter wavelengths and lower energy with longer wavelengths. Which waves above would have the highest energy, also being the most dangerous?

2) Which waves above have the lowest energy and are the most harmless?

3) Where is visible light on this spectrum?

4) Why do you think cell phones use microwaves to communicate?

5) Longer wavelengths can travel around corners easier. Which would travel farther, radio waves or microwaves?

6) Which would you see more of radio towers or cell phone towers? Explain why.

7) Why do you think remote controls use infrared waves to control TVs, drones, and remote control cars?

8) How do you think a pilot in Arizona controls a drone in the Middle East?

9) Why do you think the doctor puts a lead shield over you when you get an x-ray?

10) What do you think will give you a more detailed image, a DVD using red laser light or a DVD using blue? Explain why.

11) When using a telescope to look into outer space, which will give you more information and details, using the light waves or x-rays? Explain why.

12) Since the big Bang occurred 13.77 billion years ago, and its energy is lost over time, which category of waves would you expect to find its echo? Explain why.

13) Why do you think hospitals and restaurants use ultraviolet rays for sterilization?

14) Why do you think x-rays can see our bones?

How we use Microwaves

Directions:

You will need a **microwave oven**, two **microwave bowls**, **sand**, **water**, **oven mitts**, and a **digital thermometer** connected to an **interface** that is connected to a **computer** with **Logger Pro**. Microwaves cook food by flipping water molecules so fast that the friction creates heat. **Looking at the materials and lab we will be using, what are the safety precautions we should take to protect ourselves and materials during this investigation?**

1) Microwaves cook food by flipping water molecules so fast (about 2.45 billion times per second) that the friction creates heat. Pour dry sand into a microwave bowl and check the temperature of the sand with the temperature probe. Write this data in Data Table 1.
2) Place the dry sand into the microwave and run the microwave for one minute. Using oven mitts, take the bowl of sand out of the microwave and measure the temperature with the probe. Write this data in Data Table 1.
3) Subtract the two numbers and write down the change in temperature in Data Table 1.
4) Place some more sand into another bowl and add some water. Measure the temperature of the sand and water before putting it in the microwave. Write this in Data Table 1.
5) Put it in the microwave for one minute, and then, using your oven mitts, take the bowl of sand out of the microwave and measure the temperature with the digital thermometer. Write this data in Data Table 1.
6) Subtract the two numbers and write down the change in temperature in Data Table 1.

Data Table 1

Type of Sand	Temperature before Microwave (°C)	Temp after Microwave (°C)	Change in Temperature
Dry Sand			
Wet Sand			

Questions:

1) Which bowl had a greater temperature change?

2) Why do you think this happened?

3) How does this explain why dried food stuck to the inside of the microwave does not get hot?

4) Will a microwave always cook food? Explain.

5) Why do you think microwaves interact with water this way? What else spins with electromagnetics?

Simple Physics Investigations   Seven Sides Publishing

Making a Rainbow

Directions and Questions:

You will need a **sunny day** and a **hose** or **spray bottle** to make mist. **Looking at the materials and lab we will be using, what are the safety precautions we should take to protect ourselves and materials during the investigation?**

1) Make mist with a hose or a spray bottle with the sun behind you until you see a rainbow. Try to see the entire rainbow. What is the shape of the rainbow?

2) What is the order of the colors of the rainbow?

3) Which color is on the outside?

4) Which color is on the inside?

5) Which color do you think has the longest wavelength? Explain why.

6) Which color do you think has the shortest wavelength? Explain why.

7) AM radio can be heard across the country because the wavelength is longer and can get past the curvature of the Earth; this is not the case for FM radio; once out of town, the radio station seems to go out. Which color do you think would be able to go around the particles in the air when the light passes through the thickest part of the atmosphere? Explain why.

8) What color is the sun when we see it on the horizon (sunrise or sunset)? Explain why.

9) What color is the moon when we see it on the horizon? Explain why.

Extention: If you have some **diffraction gradient glasses**, put those on and explain what you see happen to all the light you see. These act similar to how the water diffracts the light we just saw through the rainbow.

Water Refraction

Directions:

You will need a **square tank** half-filled with **water**, an **Erlenmeyer flask** filled with **water**, and a **ruler**. **Looking at the materials and lab we will be using, what are the safety precautions we should take to protect ourselves and materials during the investigation?**

1) Put the ruler behind the tank with it up and down perpendicular to the floor. Look through the water from the front at the same level as the tank. Draw what you see below.

2) Look at the ruler through the tank from the front above the tank. Draw what you see below.

3) Turn the tank so that a corner is facing you. Put the ruler on the right-hand side; where do you see the ruler? Draw it below.

Questions:

1) Why do you think the images appear different through the same tank of water at different angles?

2) How would the image you see be different if the water's surface reflected light like a mirror instead of bending it?

3) How would the image change if the container was curved, bending out?

4) Pull out an Erlenmeyer flask filled with water. Look through it, and tell how things look different at different distances.

5) How does this explain how convex lenses work?

Test Tube Lenses

Directions:

Fill a **glass test tube** with **water** and seal it with a **rubber stopper**. Keep your finger or thumb over the rubber stopper so that the stopper does not fall off and spill water. **Looking at the materials and lab we will be using, what are the safety precautions we should take to protect ourselves and materials during the investigation?**

1) Set the test tube on the paper over the title of this lab. Write your observation in Data Table 1 about what you see through the test tube.
2) Hold the test tube approximately 1 cm over the title of this lab and observe it again. Record your observations in Data Table 1 below.
3) Repeat this three more times, increasing height approximately a centimeter each time. Write your observations in Data Table 1 below.

Data Table 1

Height	Observation of **Test Tube Lenses**
Right on the surface	
1 cm above the surface	
2 cm above the surface	
3 cm above the surface	
4 cm above the surface	

Questions:

1) Are the images you see real or virtual?

2) How high above the surface did the image become inverted?

3) What kind of lens does the test tube make (concave or convex)?

4) How does a magnifying glass compare to the test tube you looked through?

Reflection Lab

Directions:

You will need a **flat mirror**, **paper**, and a **protractor**. **Looking at the materials and lab we will be using, what are the safety precautions we should take to protect ourselves and materials during the investigation?**

1) Have your mirror facing you sitting on this paper where it says Place Mirror Here where three angles are drawn on it going into the mirror and a normal.
2) Ensure the normal is straight in and out of the mirror to properly see each of those lines pointing at you in the mirror; draw extensions for each line as if it came out of the mirror.
3) Measure the angle of incidence of the lines already there with respect to the normal.
4) Now measure the angles of reflection.

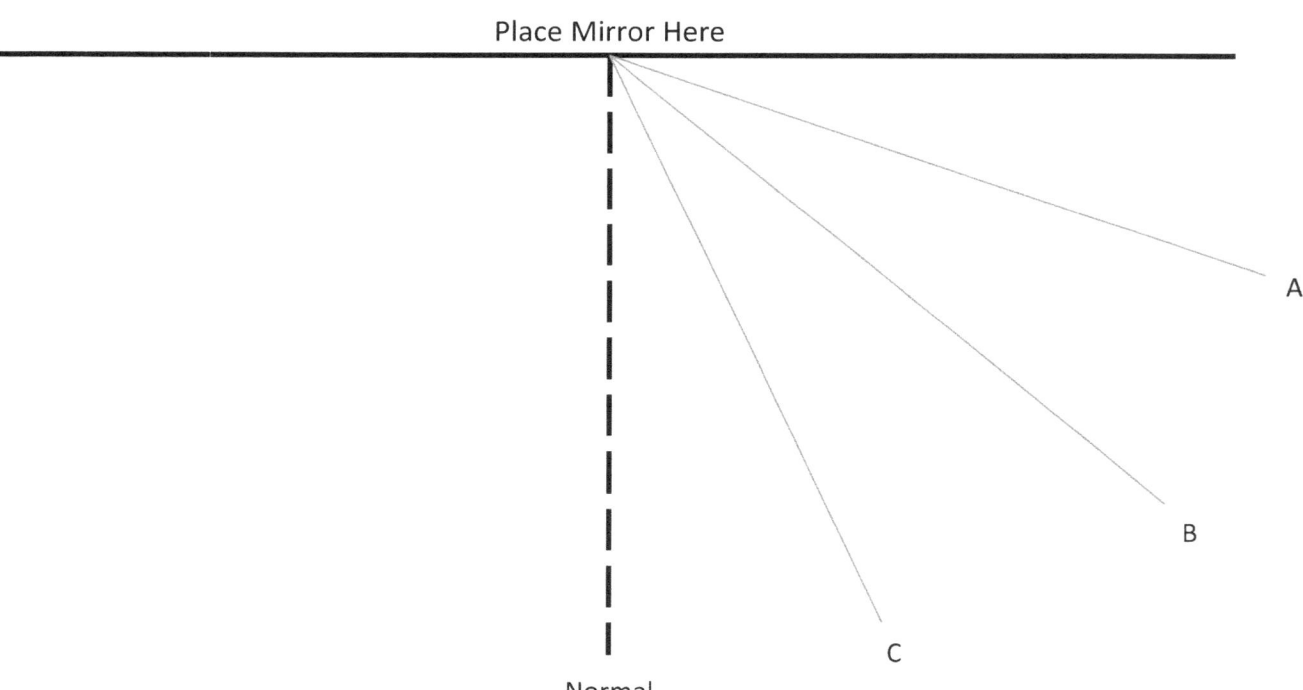

Questions:

1) What is the angle of incidence of ray A?

2) What is the angle of reflection of ray A?

3) What is the angle of incidence of ray B?

4) What is the angle of reflection of ray B?

5) What is the angle of incidence of ray C?

6) What is the angle of reflection of ray C?

7) What do you notice about the angles of the ray of reflections you drew compared to the angles of incidence?

8) Write a rule that states what you observed.

9) How would this information help you play a game of pool?

10) What is the line called where the angle of incidence and the angle of reflection is zero?

Magnifying Power

Directions:

You will need a **direct light source** (**bright lamp**, **flashlight**, or **sun**), two different **power** (thicknesses) **lenses/magnifying glass**, and a **metric ruler**. Looking at the materials and lab we will be using, what are the safety precautions we should take to protect ourselves and materials during the investigation?

1) To find the focal length of both lenses, you will want to have the light source behind/above you and any flat surface in front of you. Move your lens forward and backward until you find the smallest diameter of a light spot on the front surface in front of you. Measure the distance your lens is from that surface; this is your focal length. Write this down in Data Table 1.
2) Calculate the magnifying power using focal length in cm with the equation:

    MP = 25cm/FL where MP = Magnifying Power, FL= Focal Length. Write this number in Data Table 1 for magnifying power using focal length.
3) To find the magnifying power using the length of an image you see, draw a 1 cm line on this paper, then place the lens over the image and see it clearly and larger. Hold your metric ruler up just above the image so you can see both the ruler and the line. Measure how long the line appears in the lens in centimeters; this now tells you the magnifying power. Place the magnifying power using the length of the image for both lenses in Data Table 1.

Data Table 1

Lens	Focal Length (cm)	Magnifying Power using Focal Length	Magnifying Power using Length of Image
Thick	(cm)		
Thin	(cm)		

Questions:

1) Describe the image formed by the magnifying glass/lens.

2) Is the image you see through the lens a real or virtual image? Tell why.

3) Compare the methods of calculating Magnifying Power. How close are the results?

4) Which method do you think is most accurate? Tell why.

5) What is happening to the light as it passes through the lens to cause the image to change?

6) How does this relate to the curvature of the lens?

7) In the movie *A Bug's Life*, an ant uses a drop of water as a lens. Is it possible to use a drop of water as a lens?

Brightness and Distance

Directions:

You will need a **light source,** a **laser light**, a **meter stick**, a **light sensor** attached to an **interface** connected to a **computer** with **Logger Pro**, or a **Vernier Dynamics System and Optics Expansion Kit**. **Looking at the materials and lab we will be using, what are the safety precautions we should take to protect ourselves and materials during the investigation?**

1) In Logger Pro, open the folder Physics with Vernier and file # 29 Light Brightness Distance. Press "Collect."
2) Turn on your light source and turn off the other lights in the room. Place the 0 of the meter stick at the light source. Then place the front of the light sensor at each distance in Data Table 1 below. Measure the light intensity for each of those distances. Press "Keep" when the intensity value stabilizes at each distance. Write this data in Data Table 1 below, and press "Stop" when you have finished collecting all of the data.

Data Table 1

Distance (cm)	Intensity
5	
10	
15	
20	
25	
30	
35	
40	
45	
50	

Questions:

1) In Logger Pro, look at the graph of light intensity vs. distance. What is the relationship between distance and intensity?

2) Why do you think this happens?

3) How does this relationship help us find the distance between galaxies and stars?

4) How does this relationship relate to planets and how close they can be to stars and possibly support life (the goldilocks zone)?

5) If the orbit of the Earth is not stable, when would temperatures be higher than normal?

    a. When would it be lower than normal?

6) Lasers are focused light made by mirrors and lenses. Try to shine a laser light into the light sensor. How does that compare to the other readings?

7) Change the distance the sensor is from the laser. Did the intensity change? Why do you think that is?

Polarization of Light

Directions:

You will need a **light source**, two **polarizing filters**, and a **light sensor** attached to an **interface** connected to a **computer** with **Logger Pro. Looking at the materials and lab we will be using, what are the safety precautions we should take to protect ourselves and materials during the investigation?**

1) In Logger Pro, open the folder Physics with Vernier and the file # 28 Polarization of Light. Press "Collect."
2) Turn on your light source and turn off all the other lights in the room. Place your light sensor at a distance away from your light source and measure the light intensity. Write the data in Data Table 1 below.
3) Place a polarizing filter between the sensor and the light. Measure the light intensity and write the data in Data Table 1 below.
4) Place another polarizing filter between the sensor and the light in a way that still allows light to pass through. Measure the light intensity and write the data in Data Table 1 below.
5) Rotate one filter 90° to block the light from the light source going to the sensor. Measure the light intensity now and write that data in Data Table 1 below.

Data Table 1

Setup	Light Intensity
Just light	
Light through a filter	
Light through 2 filters	
Light through 2 filters at 90°	

Questions:

1) How did the light intensity change when the filter was placed in front of the light?

2) How did the light intensity change when two filters were placed in front of the light?

3) What happened to the light intensity when one lens was rotated 90°?

4) Hold the two filters up to the light and rotate one. What do you see?

5) Why do you think this happens?

6) Polarized sunglasses do the same thing. How can you tell if a pair of sunglasses are really polarized?

3D Glasses

Directions and Questions:

You will need two pairs of **3D glasses. Looking at the materials and lab we will be using, what are the safety precautions we should take to protect ourselves and materials during the investigation?**

1) Face the glasses towards each other directly in front of each other. What color do you see through the lenses when looking through both sets of lenses simultaneously?

2) Move the glasses closest to you back and forth from left to right. How do the colors change when looking through both sets of lenses on the glasses simultaneously?

3) Turn one set of glasses perpendicular to the ground. How did the color change as you looked through both sets of lenses?

4) Face the glasses in the same direction (looking at them from behind) and turn one pair perpendicular to the ground; how does the lens color change when looking through both sets of lenses of the glasses at the same time?

5) Put one pair of glasses on and look at the other with both eyes open. How do the lenses look on the other pair of glasses?

6) Close your right eye; what do you see now in the other glasses?

7) Close your left eye; what do you see now in the other glasses?

8) Polarized sunglasses block the horizontal light from hitting your eyes and allow the vertical light through; this keeps you from seeing glare from the sun reflecting off surfaces. How could you tell if lenses at the sunglasses store are really polarized?

9) The slits on the 3D glasses have one lens with microscopic slits that block light coming at you vertically, whereas the other lens blocks the light coming at you horizontally. If they are turned at right angles to each other, it blocks all light. Two projectors show the same movie, just staggered with different lights so that your right eye sees one projector, and your left eye sees the other; this allows you to see a 3D image on the screen. How could this work for a TV?

Light Pipes

Directions and Questions:

You will need **fiber optics** to observe some **light sources of different colors**. Looking at the materials and lab we will be using, what are the safety precautions we should take to protect ourselves and materials during the investigation?

1) Shine a light on the side of the fiber optics. Can you see the light coming out the ends?

2) Shine light into the end of a fiber optic wire. Do you see the light coming out the other end?

   a. Can you see the light coming out the sides?

3) Keep shining the light through one end of the fiber optic and bend the fiber optic. Can you still see the light coming out the other end?

   a. Can you see it coming out the sides?

4) Try shining the light on the other end. Does light still come out at the opposite end?

5) Why do you think they call these light pipes?

6) How can they be used to send information from one computer to another to have the computers talk to each other?

7) Try shining a different color of light down one end of the pipe. What do you see?

Electron Basics

Directions:

Use what you have learned to answer these questions. If you do not know how to answer the question, use the **internet** to help you.

1) How does an electron move to a higher energy level?

2) How does an electron move to a lower energy level?

3) What is the relationship between a photon and an electron?

4) What is the duality of light?

   a. This duality is the basis for String Theory, showing all particles are vibrating strings or waves. How does this show $E=mc^2$?

5) What is given off when atoms hit each other so hard the electrons fall off (why does plasma glow)?

6) How is plasma different than gas?

7) What is the electron cloud? Describe it.

8) Why do you think electrons appear everywhere in the electron cloud at the same time?

9) How does the atom seem solid when it is really empty space?

Simple Physics Investigations — Seven Sides Publishing

Build a Solar Oven

Directions:

Build a solar oven using a **box**, **aluminum foil**, **plastic wrap**, and a **black plate** to collect light and build up thermal energy to cook food of your choice. Your teacher can have some wild card items available to add to improve the efficiency of your oven. Write out how the photons' energy was collected and used to cook your food below.

Wave Technology

Directions:

Use the **internet** and your **textbook** to research how each technology below uses wave behavior and interactions with matter to capture, transmit, store, or interpret energy and information.

Solar Cells:

Medical Imaging:

Communication Technology:

Questions:

1) How are waves able to be digitized, stored, and transferred?

2) What are the advantages of using digital technology to transmit and store information?

3) What are the disadvantages of using digital technology to transmit and store information? Give examples of real-life problems.

Virtual Investigations that go with Light

ExploreLearning.com:

    Basic Prism Gizmo

    Refraction Gizmo

    Laser Reflection Gizmo

    Ray Tracing (Mirrors) Gizmo

    Photoelectric Effect Gizmo

    Star Spectra Gizmo

    Additive Colors Gizmo

    Subtractive Colors Gizmo

    Color Absorption Gizmo

    Herschel Experiment Gizmo

    Penumbra Effect Gizmo

    Bohr Model of Hydrogen Atom Gizmo

    Big Bang Theory – Hubble's Law Gizmo

PhET.colorado.com:

    Band Structure

    Bending Light

    Blackbody Spectrum

    Color Vision

    Davisson-Germer: Electron Diffraction

    Fourier: Making Waves

    Geometric Optics

    Lasers

Microwaves

Molecules and Light

Neon Lights and Other Discharge Lamps

Optical Quantum Control

Photoelectric Effect

Quantum Wave interference

Radiating Charge

Radio waves and Electromagnetic Fields

Simplified MRI

Physicsclassroom.com

Physics Interactives:

Light and Color

Electromagnetic Spectrum Infographic

RGB Addition

Paint with CMY

Color Shadows

Filtering Away

Colored Filters

Stage Lighting

Viewed in Another Light

Young's Experiment

Reflection and Refraction

Plane Mirror Images

Who Can See Who?

Optics Bench – Mirrors

Name that Image (Mirror Version)

Concave Mirror Image Formation

Convex Mirror Image Formation

Refraction

Least Time Principle

Optics Bench – Lenses

Converging Lens Image Formation

Diverging Lens Image Formation

Concept Builders:

Light and Color

Spectrum

Light Intensity

Color Addition and Subtraction

If This. Then That: Color Subtraction

Color Pigments

Color Filters

Reflection and Refraction

Law of Reflection

Who Can See Who?

The L*O*S*T Art of Image Description (Curved Mirrors)

The L*O*S*T Art of Image Description (Converging Lenses)

Law Enforcement: Refraction

Total Internal Reflection

Unit 14 Nuclear

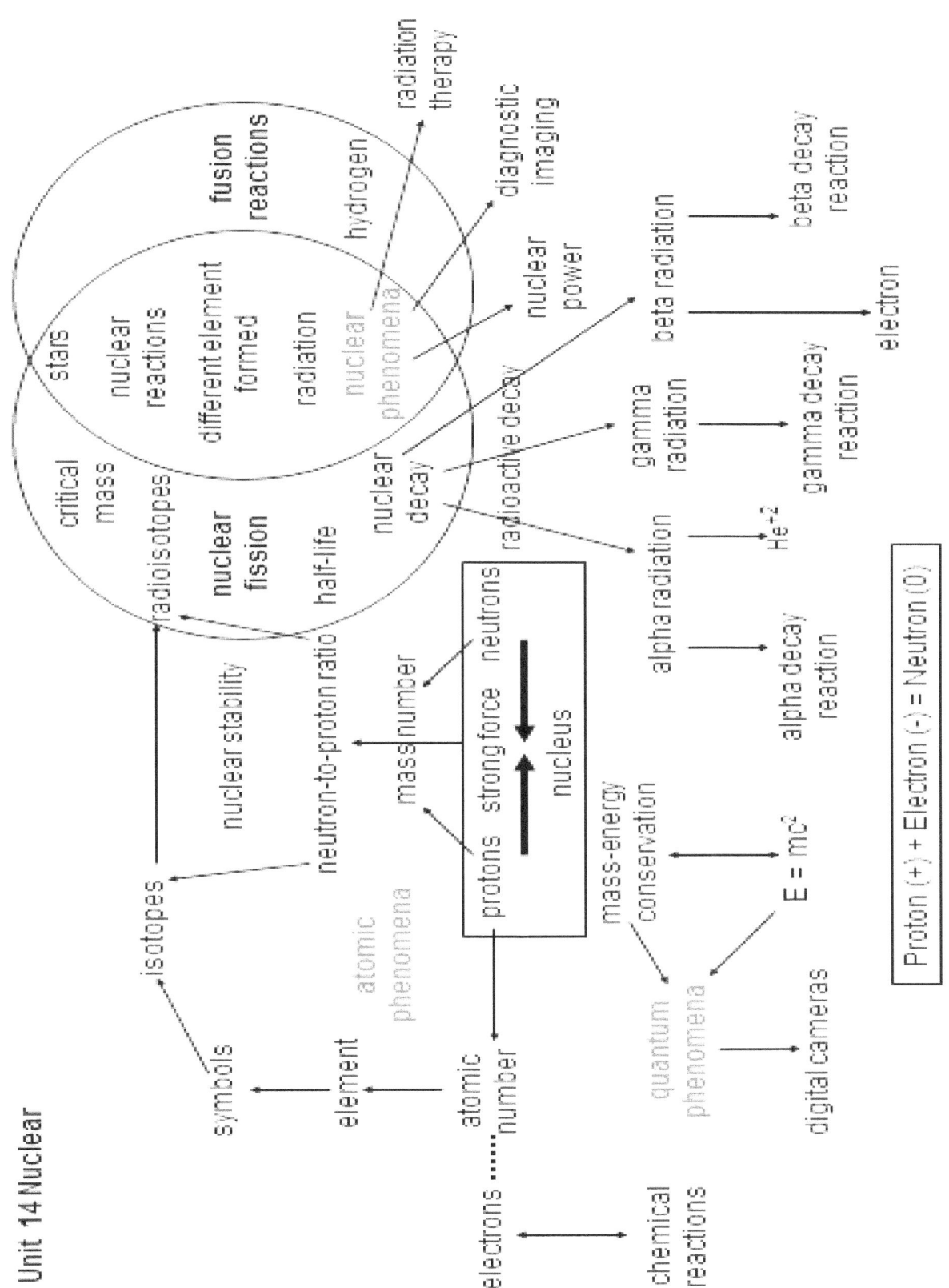

Simple Physics Investigations — Seven Sides Publishing

Scale Model of a Hydrogen Atom

Directions and Questions:

You will need a **golf ball**, a **bead**, and a **large field** or **parking lot**. **Looking at the materials and lab we will be using, what are the safety precautions we should take to protect ourselves and materials during the investigation?**

1) Walk out to a large field or parking lot, at least the size of a football field. Keep in mind the space you use still may be too small to be a scale model. You will make a model of a hydrogen atom with 1 proton and 1 electron.
2) On one edge, take a small red bead representing an electron and put that somewhere where you can see it (hang it on a fence or a tiny branch).
3) Walk at about 100 yards away; if you have more room, you can use that. Hold up the golf ball, which is a proton, read the information, and answer the questions that follow. Can you see the bead?

4) This distance is how far away the closest electron speeds around the proton. The speed approaches the speed of light. In the atom, the electron moves so fast that it makes a ball the size of a football stadium if we use these materials in the model. If you have ever seen a fan moving really fast, does it look like a disk? But is it a disk?

5) So we have to ask ourselves, the atom looks like a ball, but is it a ball? Explain why.

6) When an object moves close to the speed of light, time stops for that object. Quantum physics allows objects without time to be anywhere they would normally be at the same time because it is without time. So why is the electron said to be everywhere in the electron cloud at the same time?

7) What do you see between the proton and electron?

8) What would happen to the electron if time were to stop?

9) If that would happen to the electron, what would happen to the atom?

10) The proton is structured similarly to this. Quarks spin and orbit inside the proton close to the speed of light, just like the electron orbits around the atom. Knowing this, how does the atom show $E=mc^2$?

11) What would happen to all atoms in the area where time stops?

12) This relationship is what we call the space-time continuum. There cannot be space without time. Black holes form this way. When time disappears, so does space. What color is a black hole? Explain why.

13) Is there any space at the bottom of the black hole? Explain.

14) What do you think the temperature is at the bottom of the black hole?

   a. What is the volume of gas at this temperature?

15) This model also shows that the pixel of our universe is an illusion. If the pixel of our universe is an illusion, what does that say about our universe?

16) What do you think is the ultimate reality?

17) With our laws of physics, are we allowed to know?

Half-life of Pennies

Directions:

You will need a **plastic tub** (about the size of a shoebox) with a **lid** and 100 **pennies. Looking at the materials and lab we will be using, what are the safety precautions we should take to protect ourselves and materials during the investigation?**

1) A radioactive element's half-life is how long it takes for half the atoms to change into a different but stable element as they go through radioactive decay. Since pennies have two sides, almost half will land on heads when flipped, and almost half will land on tails.
2) In your tub, place all 100 pennies heads up. These will represent your radioactive isotopes.
3) Place the lid on your box, shake it and count to 10.
4) Lift the lid and take out all the pennies that have landed tails up. Count them and fill in Data Table 1. Subtract the number of tails from 100 to show the number still heads up.
5) Now only the pennies that are heads up are still in the tub. Repeat the procedure in #s 3-4 six more times unless all your pennies turned up tails before that.
6) Graph your data from Data Table 1 on Graph 1
7) Your teacher will give each group a number and write your data into Data Table 2 for your group number. Get the other data for the other groups and write them in Data Table 2
8) Average each half-life for all the groups by adding up their data and dividing by the number of groups.
9) Graph the averaged data you have for the class in Data Table 2 on Graph 2.

Data Table 1

Shaking Time (s)	# of Heads	# of Tails taken out
10		
20		
30		
40		
50		
60		
70		

Data Table 2

Groups	# of Heads at 0 (s)	# of Heads at 10 (s)	# of Heads at 20 (s)	# of Heads at 30 (s)	# of Heads at 40 (s)	# of Heads at 50 (s)	# of Heads at 60 (s)	# of Heads at 70 (s)
1	100							
2	100							
3	100							
4	100							
5	100							
6	100							
7	100							
8	100							
9	100							
10	100							
11	100							
12	100							
13	100							
14	100							
15	100							
Average	100							

Graph 1

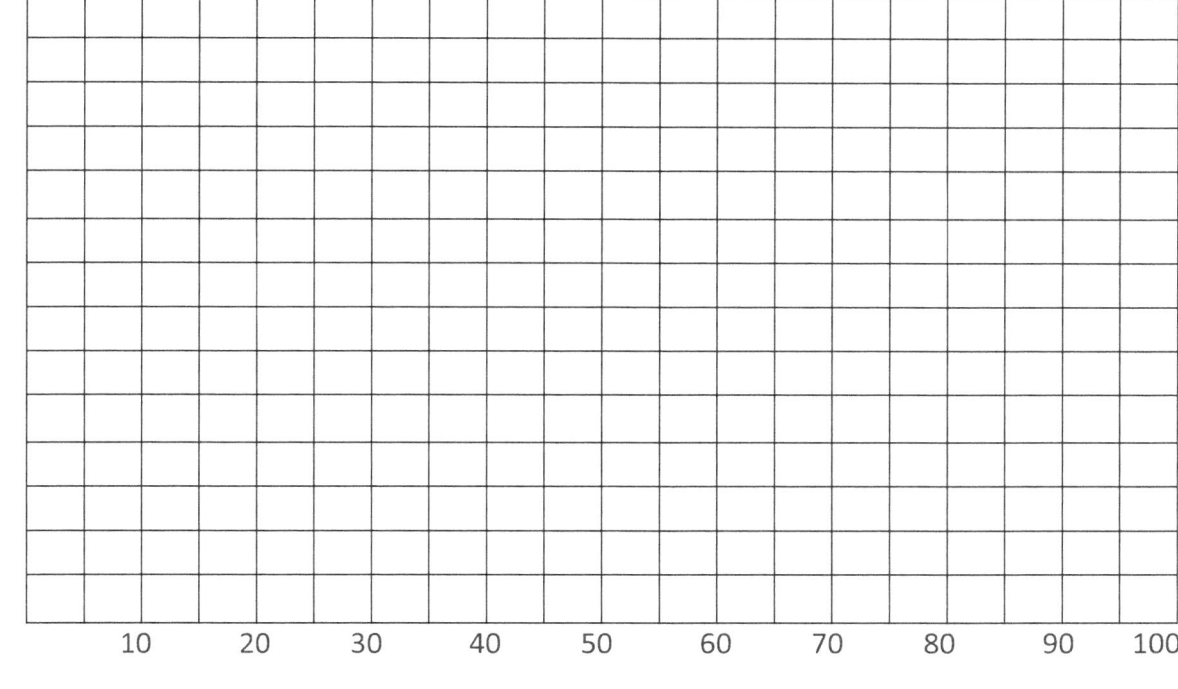

Graph 2

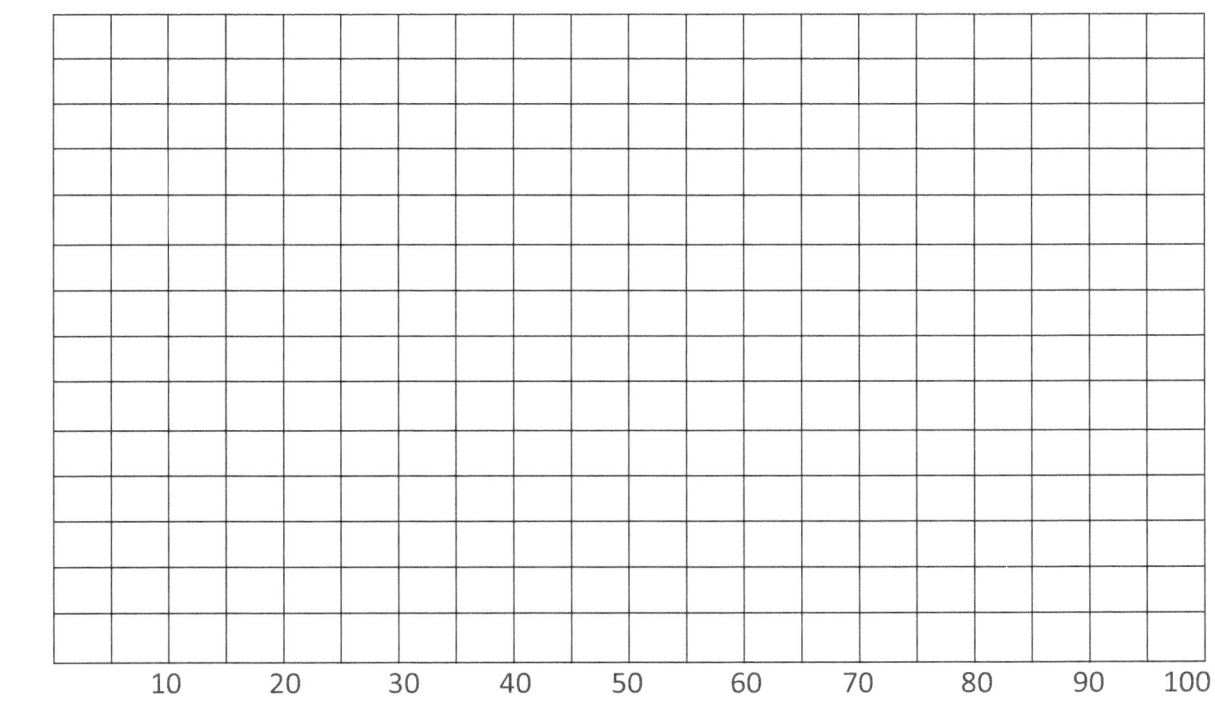

Questions:

1) What do the 10 seconds represent?

2) Which side of the coin was a stable isotope?

3) Which side of the coin was the unstable isotope?

4) What represented the radioactive decay?

5) How much of the atoms decay during each half-life?

6) How many half-lives until you should run out of pennies if all things go as expected?

7) If you start with 100 pennies and a half-life goes by every 10 seconds, how many pennies should there be after 40 seconds?

   a. How close was this to your group's results?

   b. How close was this to the class results?

   c. Why may the class's results be closer to the expected results?

Simple Physics Investigations

Calculating Nuclear Half-life Decay

Directions and Questions:

Use **colored pencils** to color in the graph below as you follow the directions to simulate what half-life looks like.

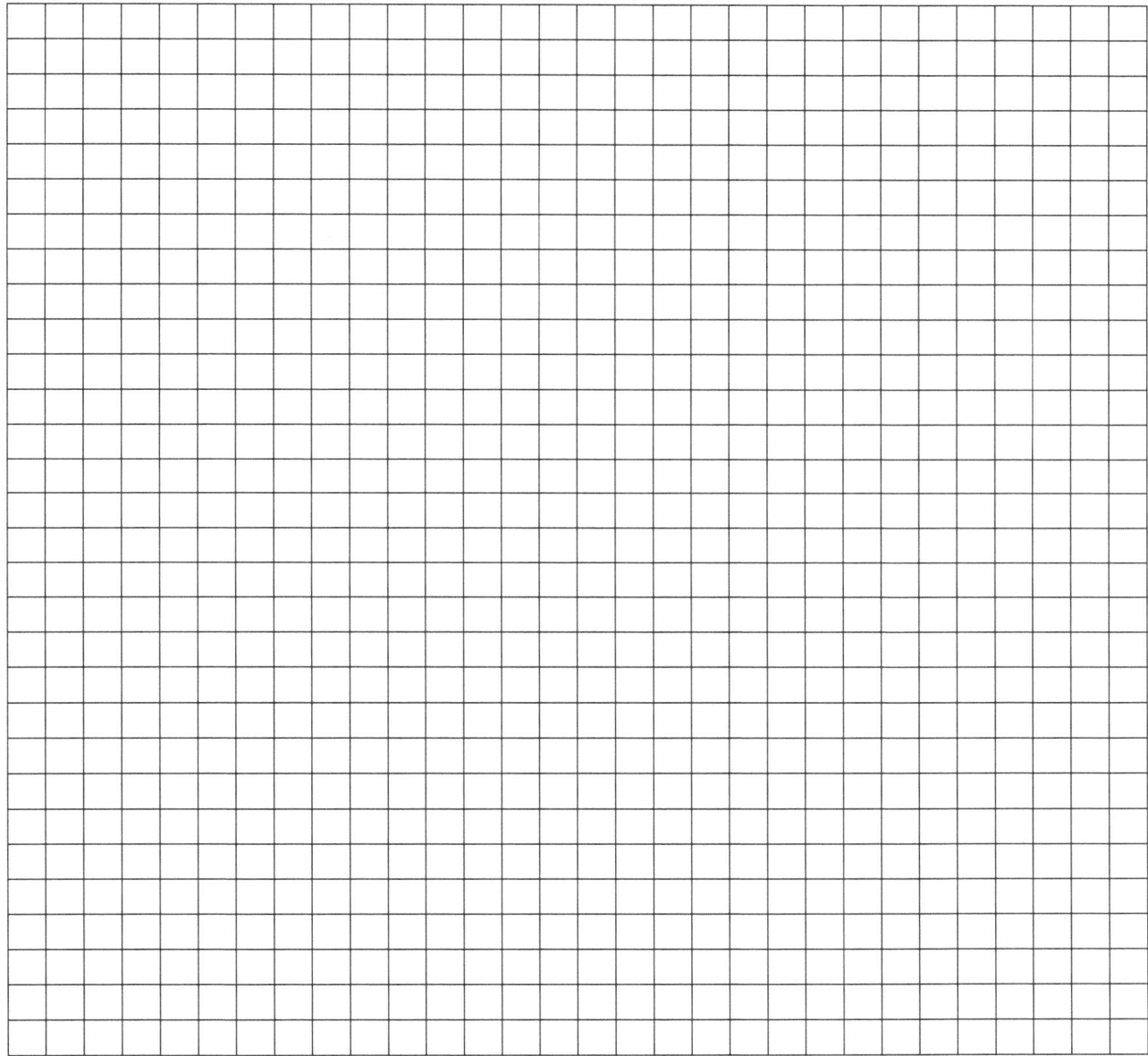

1) There are 900 squares above. Use a red colored pencil to shade in half of the squares. How many squares are left after one half-life?

2) Use a blue colored pencil to shade in half of the squares that are left; how many squares are left after two half-lives?

3) Use a green colored pencil and color in half of the squares now left. How many squares are left after three half-lives?

4) Use a yellow colored pencil and color in half of the squares now left. How many squares are left after four half-lives?

5) Use a purple colored pencil and color in half of the squares now left. How many squares are left after five half-lives?

6) Use a brown colored pencil and color in half of the squares now left. How many squares are left after six half-lives?

7) Use a black colored pencil and color in half of the squares now left. How many squares are left after seven half-lives?

8) Suppose each square represented one atom of a substance that is decomposing. How many half-lives could go by until we should expect all the atoms to be gone? Explain why.

9) If each half-life was five days, how long would it take for there to be 14 squares left?

10) If each half-life was 500 years, and there were 225 squares left, how much time went by?

Nuclear Isotopes

Directions and Questions:

Use the **internet** and your **textbook** to research how nuclear isotopes give off nuclear radiation, describe the characteristics of nuclear radiation, and show how we use it in different ways.

1) Why do some element's isotopes give off nuclear radiation?

2) What are alpha particles, and where do they come from (make sure to show a balanced reaction)?

3) What are beta particles, and where do they come from (make sure to show a balanced reaction)?

4) What are gamma rays, and where do they come from (make sure to show a balanced reaction)?

5) Which of these three is the most dangerous to life? Explain why.

6) What is radioactive dating, and how accurate is it?

7) Which radioactive isotope do we use to determine the age of a dead organism?

   a. What is its half-life, and how long can we use it after the organism dies?

8) Which radioactive isotopes help us find the age of older fossils?

9) How old is the oldest rock scientists have found?

10) What does that tell us about the age of the Earth?

11) How do we use nuclear phenomena in medicine?

   a. Would these work if the atoms were solid, or can they only happen because the atoms are mostly empty space?

12) How do we use nuclear phenomena to make electricity?

13) Draw a diagram of how a solar cell works.

Nuclear Fission and Fusion

Directions and Questions:

Use the **internet** and your **textbook** to research and explain how nuclear fission and fusion happen.

1) Draw a diagram of nuclear fission.

2) What are the element isotopes involved?

3) How does the reaction get started (what goes into the reaction)?

4) What comes out of the fission reaction?

5) How does this lead to a chain reaction?

6) How do we slow this reaction down?

7) Draw a diagram of nuclear fusion.

8) What are the element isotopes involved?

9) How is fusion different than fission?

10) What goes into the fusion reaction?

11) What comes out of the fusion reaction?

12) Which one (fission or fusion) has a waste product that is harmful to the environment?

13) Which one (fission or fusion) does not have any harmful waste products?

14) Which one (fission or fusion) do we know how to use to make usable energy?

15) Which one (fission or fusion) is being researched to make clean energy but not usable yet?

Nuclear Chain Reactions

Directions:

You will need a large number of **dominos** and a **stopwatch**. **Looking at the materials and lab we will be using, what are the safety precautions we should take to protect ourselves and materials during the investigation?**

1) Take half the dominos and set them up in one line so that when you knock the first one over, the first domino will hit the second, the second will hit the third, and so on until they all fall; this is Domino Set 1. Set 1 is like a nuclear chain reaction in a nuclear reactor.
2) Take the other set of dominos and set them up so that the first domino will knock down two dominos, and those two dominos will knock down 4, and those four will knock down 8, and so on until they are all used up; this is Domino Set 2. Set 2 is like a nuclear bomb blowing up.
3) Take your stopwatch and time how long it takes to knock down the dominos in set 1. How long did it take to knock down domino set 1?
4) Then find out how long it takes to knock down the dominos in set 2. How long did it take to knock down domino set 2?

Questions:

1) Which reaction took longer?

2) Why do you think we want nuclear chain reactions to go slower in a nuclear reactor that produces electricity?

3) What would happen if the nuclear reactor acted more like Domino Set 2?

4) How do you think we slow down this reaction and cool it off so it does not blow up?

Nuclear Reactor

Directions and Questions:

Use the **internet** and your **textbook** to research how nuclear reactors work.

1) Draw a diagram of the main parts of a nuclear reactor showing how it works.

2) Where does the reaction take place in the reactor?

3) How is the water there then used?

4) How does this make electricity?

5) How do we cool off the reactor so it does not explode?

6) How do we slow the reaction down so it does not explode?

7) What is the fallout from a nuclear reactor meltdown?

8) What do we do with the waste products?

9) Why does the US prohibit the production of any new nuclear power plants?

10) How do other countries dispose of their nuclear waste?

11) What are the positive and negative effects of using nuclear power plants to produce electricity?

12) Why do we not want developing countries to develop this technology?

Smoke Detector

Directions and Questions:

Use the **internet** and your **textbook** to research and explain how a smoke detector works with nuclear radiation.

1) Draw a diagram of how the radioactive isotopes work with the gases in the air to complete a circuit in the smoke detector.

2) Which elements are involved?

3) How are the elements interacting?

4) What happens when smoke gets in the way?

All of Physics Found in the Atom

Directions:

1) Go back and look at all the concept maps from each unit and find as many concepts as you can that are also seen in the atom. List the concepts, define what they are, and tell how they are in the atom.
2) Use whatever electronic media your teacher chooses to present your findings.
3) You must use some from each unit. The person with the most described correctly will get a 100%, and the rest will be graded based on that total.

Virtual Investigations that go with Nuclear

ExploreLearning.com:

    Element Builder Gizmo

    Nuclear Decay Gizmo

    Half-life Gizmo

    Isotopes Gizmo

    Average Atomic Mass Gizmo

PhET.colorado.edu:

    Alpha Decay

    Atomic Interactions

    Beta Decay

    Build an Atom

    Models of the Hydrogen Atom

    Nuclear Fission

    Radioactive Dating Game

    Rutherford Scattering

Physicsclassroom.com:

    Concept Builders:

        Chemistry

        Nuclear Decay

Physics and IPC TEKS and NGSS Correlations

Nature of Science Concept Map Phy c 1ABH 2AD 3AB 4AB, IPC c 1ABH 2AD 3AB 4AB

Focus on the Process Phy c 1ABCDEG 2ABD 3AB 4AB, IPC c 1ABCDEG 2ABD 3AB 4AB

Measurement Lab Phy c 1ABCDEF 2BC 3AB 4AB, IPC c 1ABCDEF 2BC 3AB 4AB

Patterns in Pennies Phy c 1ABCDEFG 2ABCD 3AB 4A, IPC c 1ABCDEFG 2ABCD 3AB 4A

Virtual Introduction Investigations Phy c 1ABEFG 2ABCD 3ABC 4AB, IPC c 1ABEFG 2ABCD 3ABC 4AB

Constant Velocity Concept Map Phy c 5ABC, IPC c 5A; HS-PS2-2

The Motion of a Toy Car Phy c 1ABCDEFG 2ABCD 3AB 4A 5ABC, IPC c 1ABCDEFG 2ABCD 3AB 4A 5A; HS-PS2-2

The Motion of a Bowling Ball Phy c 1ABCDEFG 2ABCD 3AB 4A 5ABC, IPC c 1ABCDEFG 2ABCD 3AB 4A 5A; HS-PS2-2

Motion Detector Lab Phy c 1ABCDEFG 2AB 3AB 4A 5ABC, IPC c 1ABCDEFG 2AB 3AB 4A 5A; HS-PS2-2

Virtual Constant Velocity Investigations Phy c 1ABEFGH 2ABCD 3ABC 4AB 5ABC, IPC c 1ABEFGH 2ABCD 3ABC 4AB 5A; HS-PS2-2

Constant Acceleration Concept Map Phy c 5ABC, IPC c 5A; HS-PS2-1

Marbles in Motion Phy c 1ABCDEFG 2ABC 3AB 4A 5ABC, IPC c 1ABCDEG 2AB 3AB 4A 5A; HS-PS2-1

Observing Changes in Motion Phy c 1ABCE 3ABC 4A 5CF, IPC c 1ABCE 3ABC 4A 5A; HS-PS2-1

Motion in Real Life Phys c 1ABE 3ABC 4A 5CE, IPC c 1ABC 3ABC 4A 5A; HS-PS2-12

Ball Bounce Phy c 1ABCDEG 2AB 3AB 4A 5ABC, IPC c 1ABCDEG 2AB 3AB 4A 5A; HS-PS2-1

Cart on a Ramp Phy c 1ABCDEG 2AB 3AB 4A 5ABC, IPC c 1ABCDEG 2AB 3AB 4A 5A; HS-PS2-1

Virtual Constant Acceleration Investigations Phy c 1ABEFGH 2ABCD 3ABC 4AB 5ABC, IPC c 1ABEFGH 2ABCD 3ABC 4AB 5A; HS-PS2-12

Projectile Motion Concept Map Phy c 5ABCD, IPC c 5ABD; HS-PS2-1

Picket Fence Free Fall Phy c 1ABCDEFG 2ABC 3ABC 4A 5ACD, IPC c 1ABCDEFG 2ABC 3ABC 4A 5ABD; HS-PS2-1

Observing Vertical Motion of a Kickball Phy c 1ABCDEFGH 2ABCD 3ABC 4A 5ACD, IPC c 1ABCDEFGH 2ABCD 3ABC 4A 5ABD; HS-PS2-1

Measuring the Effects of Air Resistance Phy c 1ABCDEFG 2ABC 3AB 4A 5AC, IPC c 1ABCDEFG 2ABC 3AB 4A 5ABD; HS-PS2-12

Air Resistance Phy c 1ABCDEFG 2ABC 3AB 4A 5AC, IPC c 1ABCDEFG 2ABC 3AB 4A 5ABD; HS-PS2-12

Elevator Lab Phy c 1ABCDE 2B 3AB 4A 5AC, IPC c 1ABCDE 2B 3AB 4A 5ABD; HS-PS2-12

Shoot and Drop Phy c 1ABCDE 3AB 4A 5CE; HS-PS2-12

Ball and Cart Phy c 1ABCDE 3AB 4A 5C; HS-PS2-12

Projectile Motion Phy c 1ABCDEFH 2ABC 3AB 4A 5ABCD, IPC c 1ABCDEFH 2ABC 3AB 4A 5ABD; HS-PS2-12

Virtual Projectile Investigations Phy c 1ABEFGH 2ABCD 3ABC 4AB 5ABCD, IPC c 1ABEFGH 2ABCD 3ABC 4AB 5ABD; HS-PS2-12

Forces Concept Map Phy c 5BCEFG, IPC c 5BD; HS-PS2-1

The Human Table Phy c 1ABC 3ABC 4A 5F, IPC c 1ABC 3ABC 4A 5BD; HS-PS2-12

Balancing Forks Phy c 1ABCD 3ABC 4A 5F, IPC c 1ABCD 3ABC 4A 5BD; HS-PS2-12

Weight isn't Mass Lab Phy c 1ABCDEFG 2AB 3AB 4A 5CDEF, IPC c 1ABCDEFG 2AB 3AB 4A 5BD; HS-PS2-1

Comparing Friction Phy c 1ABCDEF 2AB 3AB 4A 5ABEF, IPC c 1ABCDEF 2AB 3AB 4A 5B; HS-PS2-12

Friction Lab Phy c 1ABCDEF 2AB 3AB 4A 5ABEF, IPC c 1ABCDEF 2AB 3AB 4A 5B; HS-PS2-12

Newton's Relay Race Phy c 1ABCDE 3AB 4A 5ABCEF, IPC c ABCDE 3AB 4A 5B; HS-PS2-12

Observing Inertia Newton's First Law of Motion Phy c 1ABCDEGH 3AB 4A 5E, IPC c 1ABCDEGH 3AB 4A 5C; HS-PS2-2

Inertia Lab Stations Phy c 1ABCEG 3AB 4A 5CE, IPC c 1ABCE 3AB 4A G5BC; HS-PS2-2

Newton's Second Law Phy c 1ABCDEFGH 2AB 3AB 4A 5ABCEF, IPC c 1ABCDEFGH 2AB 3AB 4A 5B; HS-PS2-1

Fan Cart Lab Phy c 1ABCDEFH 2BC 3AB 4A 5ACEF, IPC c 1ABCDEFH 2BC 3AB 4A 5B; HS-PS2-1

Newton's Third Law Phy c 1ABCDEFGH 2AB 3AB 4A 5ABF, IPC c 1ABCDEFGH 2AB 3AB 4A 5B; HS-PS2-2

Water Bottle Rockets Phy c 1ABCDE 2BCD 3AB 4A 5BCDEF, IPC c 1ABCDE 2BCD 3AB 4A 5ABD; HS-PS2-12

Virtual Force Investigations Phy c 1ABEFGH 2ABCD 3AB 4AB 5ABCDEF, IPC c 1ABEFGH 2ABCD 3AB 4AB 5ABCD; HS-PS2-12

Gravitational and Centripetal Forces Concept Map Phy c 1H 5AB, IPC c 4DF 5ABD; HS-PS2-124

Bending of Space-time Phy c 1ABCDE 2ABD 3AB 4D 5AB, IPC c 1ABCDE 2ABD 3AB 4DF 5AB; HS-PS2-4

The Push of Gravity on Earth Phy c 1ABEH 2BC 3ABC 4AC 5AB, IPC c 1ABEH 2BC 3ABC 4DF 5BD; HS-PS2-4

Observing Centripetal Force Phy c 1ABDE 3ABC 4A 5AB, IPC c 1ABDE 3ABC 4DF 5AB; HS-PS2-124

Centripetal Force Under Glass Phy c 1ABCDEH 3ABC 4A 5AB, IPC c 1ABCDEH 3ABC 4DF 5AB; HS-PS2-12

Uniform Circular Motion Lab Phy c 1ABCDEFH 2ABC 3AB 4A 5AB, IPC c 1ABCDEFH 2ABC 3AB 4DF 5AB; HS-PS2-12

Simulating the Orbit of a Planet and Sun Phy c 1ABCDE 2B 3AB 4A 5AB, IPC c 1ABCDE 2B 3AB 4DF 5ABD; HS-PS2-124

Observing Forces in Orbit Phy c 1ABCDE 2ABD 3AB 4A 5AB, IPC c 1ABCDE 2ABD 3AB 4DF 5ABD; HS-PS2-124

Virtual Gravitational and Centripetal Forces Investigations Phy c 1ABEFGH 2ABCD 3ABC 4AB 5AB, IPC c 1ABEFGH 2ABCD 3ABC 4AB 5AB; HS-PS2-124

Momentum Concept Map Phy c 7DE, IPC c 5C 6C; HS-PS3-12

Velocity and Momentum Phy c 1ABCDEFG 2BC 3AB 4A 7DE, IPC c 1ABCDEFG 2BC 3AB 4A 5C 6C; HS-PS3-12

The Momentum of Colliding Objects Phy c 1ABCDEF 2BC 3AB 4A 7DE, IPC c 1ABCDEF 2BC 3AB 4A 5C 6C; HS-PS3-2

Conservation of Momentum Phy c 1ABCDEF 2ABCD 3AB 4A 7DE, IPC c 1ABCDEF 2ABCD 3AB 4A 5C 6C; HS-PS3-12

Egg Drop Phy c 1ABCDEG 2ABD 3AB 4A 7ABCDE, IPC c 1ABCDEG 2ABD 3AB 4A 5C; HS-PS2-3, 3-2

Virtual Momentum Investigations Phy c 1ABEFGH 2ABCD 3ABC 4AB 7ABCDE, IPC c 1ABEFGH 2ABCD 3ABC 4AB 5C 6CD; HS-PS2-3, 3-12

Energy, Work, and Power Concept Map Phy c 7ABC 8A, IPC c 6CD; HS-PS3-12

The Energy of a Pendulum Lab Phy c 1ABCDEF 2B 3ABC 4A 7BC 8A, IPC c 1ABCDEF 2B 3ABC 4A 6CD; HS-PS3-12

Potential and Kinetic Energy Phy c 1ABCDEFG 2ABC 3AB 4A 5CDEF 7BC, IPC c 1ABCDEFG 2ABC 3AB 4A 6CD; HS-PS3-12

Analyzing Elastic Potential Energy Phy c 1ABCDEF 2B 3AB 4A 7BC, IPC c 1ABCDEF 2B 3AB 4A 6CD; HS-PS3-2

Happy and Sad Balls Phy c 1ABCDEF 2ABC 3ABC 4A 5CDEF 7BC, IPC c 1ABCDEF 2ABC 3ABC 4A 6CD; HS-PS3-12

Conservation of Energy in a Toy Phy c 1ABCDE 2B 3AB 4A 7C, IPC c 1ABCDE 2B 3AB 4A 6CD; HS-PS3-2

Energy and Rockets Lab Phy c 1ABCDEH 2B 3AB 4A 5CDEF 7BC, IPC c 1ABCDEH 2B 3AB 4A 6CD; HS-PS3-123-12

Who's got the Power? Phy c 1ABCDEF 2B 3AB 4A 7A, IPC c 1ABCDEF 2B 3AB 4A 6C; HS-PS3-12

Levers Lab Phy c 1ABCDEF 2BC 3AB 4A 5F 7A, IPC c 1ABCDEF 2BC 3AB 4A 6C; HS-PS3-12

Simple Machines Lab Phy c 1ABCDE 2B 3AB 4A 5F 7A, IPC c 1ABCDE 2B 3AB 4A 6C; HS-PS3-2

Pulley Lab Phy c 1ABCDE 2B 3AB 4A 5F 7A, IPC c 1ABCDE 2B 3AB 4A 6C; HS-PS3-2

Bicycle Lab Phy c 1ABCDEF 2B 3AB 4A 5CDF 7A, IPC c 1ABCDEF 2B 3AB 4A 6C; HS-PS3-2

Rube Golberg Machine Phy c 1ABCDEG 2ABD 3AB 4A 7ABC, IPC c 1ABCDEG 4ABD 3AB 4A 6C; HS-PS3-3

Virtual Energy, Work, and Power Investigations Phy c 1ABEFGH 2ABCD 3AB 4AB 5F 7ABCD, IPC c 1ABEFGH 2ABCD 3AB 4AB 6CD; HS-PS3-123

Thermodynamics Concept Map Phy c 7AC, IPC c 6CDE; HS-PS3-4

Energy Transformation Balls Phy c 1ABCDEH 3A 4A 7C, IPC c 1ABCDEH 3A 4A 6CD; HS-PS3-24

Student Atomic Motion Phy c 1ABCG 2AD 3AB 4A 7C, IPC c 1ABCG 2AD 3AB 4A 6D; HS-PS3-4

Observing Molecular Motion Phy c 1ABCDEF 3AB 4A 7C, IPC c 1ABCDEF 3AB 4A 6D; HS-PS3-4

Thermal Energy Ventilation Phy c 1ABCDEF 2B 3AB 4A 7C, IPC c 1ABCDEF 2B 3AB 4A 6D; HS-PS3-24

The Direction Thermal Energy Moves Phy c 1ABCDG 2BD 3AB, IPC c 1ABCDG 2BD 3AB 6E; HS-PS3-4

Convection in Liquids and Gases Phy c 1ABCDE 3AB 4A 7C, IPC c 1ABCDE 3AB 4A 6D; HS-PS3-24

Observing Conduction Convection and Radiation Phy c 1ABCDE 3AB 4A 7C, IPC c 1ABCDE 3AB 4A 6D; HS-PS3-24

Observing Boyles Law Phy c 1ABCDE 2B 3AB 4A 7A, IPC c 1ABCDE 2B 3AB 4A 6E; HS-PS3-4

Relationship Between Temperature, Volume, and Pressure: Charles and Gay-Lussac's Law Phy c 1ABCDE 2B 3AB 4A 7AC, IPC c 1ABCDE 2B 3AB 4A 6DE; HS-PS3-4

Testing the Rate of Heat Movement Phy c 1ABCDEF 2AB 3AB 4A 7C, IPC c 1ABCDEF 2AB 3AB 4A 6CD; HS-PS3-4

Virtual Thermodynamics Investigations Phy c 1ABEFGH 2ABCD 3AB 4AB 7AC, IPC c 1ABEFGH 2ABCD 3AB 4AB 6CDE; HS-PS3-24

Electrostatics Concept Map Phy c 6ABC, IPC c 5DE 6B; HS-PS3-5

Static Electricity Phy c 1ABCDE 3AB 4A 6BC, IPC c 1ABCDE 3AB 4A 5DE 6B; HS-PS3-25

The Spinning Match Phy c 1ABCDE 3AB 4A 6C, IPC c 1ABCDE 3AB 4A 5DE 6B; HS-PS3-5

Charged Tape Phy c 1ABCD 3AB 4A 6C, IPC c 1ABCD 3AB 4A 5DE 6B; HS-PS3-5

Van De Graaff Phy c 1ABCDE 3AB 4A 6C, IPC c 1ABCDE 3AB 4A 5DE 6B; HS-PS3-25

Calculating Forces on Charged Particles Phy c 1ABH 2BC 3ABC 4A 5BCF 6A, IPC c 1ABH 2BC 3ABC 4C 5ADE; HS-PS2-4, 3-5

Identifying Conductors and Insulators Phy c 1ABCDEF 3AB 4A 6C, IPC c 1ABCDEF 3AB 4A 5D 6E; HS-PS3-5

Virtual Electrostatics Investigations Phy c 1ABEFGH 2ABCD 3ABC 4AB 6ABC, IPC c 1ABEFGH 2ABCD 3ABC 4AB 5DE 6B; HS-PS2-4, 3-5

Circuits Concept Map Phy c 6DE, IPC c 6AB; HS-PS3-5

Battery Power Phy c 1ABCDE 3AB 4A 6CE, IPC c 1ABCDE 3AB 4A 6AB; HS-PS3-25

Making a Graphite Light Bulb Phy c 1ABCDE 3AB 4A 6CD, IPC c 1ABCDE 3AB 4A 6AB; HS-PS3-25

Human Circuits Phy c 1ABCDEG 2B 3AB 4A 6D, IPC c 1ABCDEG 2B 3AB 4A 6AB; HS-PS3-5

Comparing Series and Parallel Circuits Phy c 1ABCDEH 3AB 4A 6DE, IPC c 1ABCDEH 3AB 4A 6AB; HS-PS3-5

Measuring Actual Voltage and Resistance Phy c 1ABCDEH 2B 3AB 4A 6DE, IPC c 1ABCDEH 2B 3AB 4A 6AB; HS-PS3-5

Hotdog Circuits Phy c 1ABCDEFH 2B 3AB 4A 6DE, IPC c 1ABCDEFH 2B 3AB 4A 6AB; HS-PS3-25

Virtual Circuits Investigations Phy c 1ABEFGH 2ABCD 3AB 4AB 6CDE, IPC c 1ABEFGH 2ABCD 3AB 4AB 6AB; HS-PS3-5

Magnetism Concept Map Phy c 6B, IPC c 6B; HS-PS3-5

Seeing Magnets Phy c 1ABCDE 3AB 4A 6B, IPC c 1ABCDE 3AB 4A 6B; HS-PS3-5

Making Electromagnets Phy c 1ABCDE 2B 3AB 4A 6BC, IPC c 1ABCDE 2B 3AB 4A 6ABD; HS-PS3-5

Motors and Generators Phy c 1ABCDEH 3AB 4A 6B, IPC c 1ABCDEH 3AB 4A 6AB; HS-PS2-5, 3-235

Virtual Magnetism Investigations Phy c 1ABEFGH 2ABC 3ABC 4AB 6ABC, IPC c 1ABEFGH 2ABC 3ABC 4AB 5DE 6B; HS-PS2-5, 3-25

Waves and Sound Concept Maps Phy c 8ACBD, IPC c 6EF; HS-PS4-13

Measuring Wave Properties Phy c 1ABCDEF 2B 3AB 4A 8ABCD, IPC c 1ABCDEF 2B 3AB 4A 6EF; HS-PS4-1

Observing Waves in a Slinky Phy c 1ABCDE 3AB 4A 8ABCD, IPC c 1ABCDE 3AB 4A 6EF; HS-PS4-1

Observing Sound Phy c 1ABCDE 3AB 4A 8AB, IPC c 1ABCDE 3AB 4A 6EF; HS-PS4-13

Coffee Can Phones Phy c 1ABCDE 3AB 4A 8AB, IPC c 1ABCDE 3AB 4A 6EF; HS-PS4-13

Music Test Tubes Phy c 1ABCDEH 2B 3AB 4A 8ABCD, IPC c 1ABCDEH 2B 3AB 4A 6F; HS-PS4-13

Singing Glasses and the Dancing Toothpick Phy c 1ABCDE 3AB 4AC 8ABCD, IPC c 1ABCDE 3AB 4AC 6EF; HS-PS4-13

Playing the Rubber Band Phy c 1ABCDEF 3AB 4A 5E 8ABCD, IPC c 1ABCDEF 3AB 4A 6EF; HS-PS4-13

Music has Patterns Phy c 1ABCDEF 2BC 3AB 4A 8ABC, IPC c 1ABCDEF 2BC 3AB 4A 6EF; HS-PS4-13

Observing the Speed of Sound Phy c 1ABCDEF 2ABC 3AB 4A 8ABCD, IPC c 1ABCDEF 2ABC 3AB 4A 6EF; HS-PS4-1

The Doppler Effect Phy c 1ABCDE 3AB 4A 8D, IPC c 1ABCDE 3AB 4A 6EF; HS-PS4-13

Virtual Waves and Sound Investigations Phy c 1ABEFGH 2ABCD 3ABC 4AB 8ABCD, IPC c 1ABEFGH 2ABCD 3ABC 4AB 6EF; HS-PS4-13

Light Concept Maps Phy c 8BCDEFG 9ABCD, IPC c 6DEF 7DE; HS-PS4-1345

Uses of the Electromagnetic Spectrum Phy c 1ABEG 3AB 4A 8CE, IPC c 1ABEG 3AB 4A 5D 6F; HS-PS4-14

How we use Microwaves Phy c 1ABCDEF 3AB 4A 6B 8E, IPC c 1ABCDEF 3AB 4A 5AD 6CDE 7C; HS-PS3-4, 4-45

Making a Rainbow Phy c 1ABCDE 3AB 4A 8CDF, IPC c 1ABCDE 3AB 4A 6EF; HS-PS4-34

Water Refraction Phy c 1ABCDE 3AB 4A 8CD, IPC c 1ABCDE 3AB 4A 6EF; HS-PS4-3

Test Tube Lenses Phy c 1ABCDEF 3AB 4A 8CDG, IPC c 1ABCDEF 3AB 4A 6EF; HS-PS4-3

Reflection Lab Phy c 1ACDEG 2B 3AB 4A 8DG, IPC c 1ABCDEF 2B 3AB 4A 6EF; HS-PS4-3

Magnifying Power Phy c 1ABCDEF 3AB 4A 8DE, IPC c 1ABCDEF 3AB 4A 6EF; HS-PS4-13

Brightness and Distance Phy c 1ABCDEF 2B 3AB 4A 8CD 9A, IPC c 1ABCDEF 2B 3AB 4A 6EF; HS-PS3-12, 4-1

Polarization of Light Phy c 1ABCDEF 2B 3AB 4A 9B, IPC c 1ABCDEF 2B 3AB 4A 6EF; HS-PS4-34

3D Glasses Phy c 1ABCDE 3AB 4A 9B, IPC c 1ABCDE 3AB 4A 6EF; HS-PS4-34

Light Pipes Phy c 1ABCDE 3AB 4A 8D, IPC c 1ABCDE 3AB 4A 6EF; HS-PS4-3

Electron Basics Phy c 1AB 3AB 4AC 9CD, IPC c 1AB 3AB 4AC 7DE; HS-PS4-3

Build a Solar Oven Phy c 1ABCDEG 2ABD 3AB 4A 8DEG, IPC c 1ABCDEG 2ABD 3AB 4A 6CE 7C; HS-PS3-3, 4-5

Wave Technology Phy c 1ABE 3AB 4A 8DEF 9A, IPC c 1ABE 3AB 4A 6F 7CE; HS-PS4-25

Virtual Light Investigations Phy c 1ABEFGH 2ABCD 3ABC 4AB 8ABCDEFG 9ABC, IPC c 1ABEFGH 2ABCD 3ABC 4AB 6EF 7DE; HS-PS4-12345

Nuclear Concept Map Phy c 9CD, IPC c 5D 7DE; HS-PS1-8

Scale Model of a Hydrogen Atom Phy c 1ABCDEH 3AB 4A 9D, IPC c 1ABCDEH 3AB 4A 5D 7E; HS-PS1-8

Half-life of Pennies Phy c 1ABCDEG 2ABC 3AB 4A, IPC c 1ABCDEG 2ABC 3AB 4A 5D; HS-PS1-8

Calculating Nuclear Half-life Decay Phy c 1ABEFG 2BC 3AB 4A, IPC c 1ABEFG 2BC 3AB 4A 8C; HS-PS1-8

Nuclear Isotopes Phy c 1AB 3AB 4AC, IPC c 1AB 3AB 4AC 5D 8BCD; HS-PS1-8, 4-5

Nuclear Fission and Fusion Phy c 1AB 3AB 4AC, IPC c 1AB 3AB 4AC 8CD; HS-PS1-8, 3-24

Nuclear Chain Reactions Phy c 1ABCDE 2B 3AB 4A, IPC c 1ABCDE 2B 3AB 4A 8CD; HS-PS1-8, 3-24

Nuclear Reactor Phy c 1AB 3AB 4AC, IPC c 1AB 3AB 4AC 5D 8CD; HS-PS1-8, 3-24

Smoke Detector Phy c 1AB 3AB 4AC, IPC c 1AB 3AB 4AC 5D 8BCD; HS-PS1-8, 3-25

All of Physics Found in the Atom Phy c 1AB 3AB 4AC 5ABCDEFGH 6ABCDE 7ABCDE 8 ABCDEFG 9ABCD, IPC c AB 3AB 4AC 5ABCD 6ABCDEFG 7ABCDEF 8ABCD; HS-PS1-8, 2-12456, 3-12345, 4-1345

Virtual Nuclear Investigations Phy c 1ABEFGH 2ABCD 3ABC 4AB 9CD, IPC c 1ABEFGH 2ABCD 3ABC 4AB 5D 7DE 8BCD; HS-PS1-18, 3-24, 4-35

Equipment List for All Investigations

If you want to be able to do all the labs in this manual, here is the list of all the equipment you will need (in order of appearance):

- Small Lego sets
- Digital scales
- Meter sticks
- Temperature probes
- 100 mL graduated cylinders
- Stopwatches
- Probe-ware interfaces
- Logger Pro software
- Computers/Laptops/Tablets
- Electric toy car
- Bowling Balls
- Motion detectors
- Hot wheels tracks
- Marbles
- Small kickballs
- Small wire baskets
- Large bouncy balls
- Vernier's Dynamics Systems
- Picket fences
- Photogates
- Ring stands and clamps
- Paper
- Coffee filters
- Hanging mass sets
- Elevator
- Shoot and Drop set-up
- Ball and cannon carts
- Steel balls
- Masking tape
- Dual range force sensors
- Ice cubes
- Rocks
- Rubber erasers
- Wooden blocks
- Aluminum foil
- Trays
- Brooms
- Basketballs
- Kid's bouncy balls
- Toy cars spring-loaded
- Pennies (lots of pennies)
- Rubber bands (different sizes)

- Ping pong balls
- Ping pong paddles
- Tennis balls
- Index cards
- Low g accelerometers
- Water bottle rocket launchers
- Air pumps
- Embroidery Hoops
- Elastic fabric
- String
- PVC pipes
- Baseballs
- Softballs
- Golf balls
- Happy and sad ball set
- Plastic tubs/boxes
- Small Toys
- Air-powered rocket launchers
- 2 lb medicine balls
- 8 lb medicine balls
- Fulcrums and podiums
- Pulleys
- Screwdrivers
- Doorknobs, both lever and round
- Bicycles
- Dowels
- Steel energy transformation balls
- Large beakers
- Water
- Refrigerator and freezer
- Hotplates
- Food coloring
- Cardboard boxes
- Incandescent light bulbs
- Work light fixtures
- Pepper
- Jiffy Pop popcorn
- Hot air popper
- Popcorn
- Microwave popcorn
- Microwave
- Large syringes
- Small Marshmallows
- Plastic bottles with a valve cap
- Small sealed syringes
- Balloons
- Nickles
- Matches

- Clear plastic cups
- Erlenmeyer flask
- Rubber stopper assemblies
- Gas pressure sensors and tubing
- Electroscopes
- Plastic combs
- Van de Graaff
- Iron golf club
- Scotch tape
- Batteries
- Battery packs
- Current Conductor
- Wires with alligator clips
- Christmas lights
- Wooden spoons
- Metal forks
- Aluminum cans
- .2-.5 mm Pencil mechanical pencil lead
- 6-volt lantern batteries
- Multi-meters
- Stripped extension cords with alligator clips
- Hotdogs
- Pickles
- Bar magnets
- Horseshoe magnets
- Iron filings in plastic cases
- BBs
- Small containers
- 3D movies glasses
- Dominos and Beads
- Thinly insulated wire
- Glass Jar and Lid
- Blue tac
- Telephone cords
- Long Slinkys
- Wire hangers
- Coffee cans
- Test tubes and stoppers
- Test tube holders
- Long tubes/boxes
- Microphone probes
- Whistling football
- Something to make mist
- Microwave oven
- Sand
- Microwave bowl
- Square small tanks
- Flat mirrors

Protractors

Flashlights/bright lamps

Different magnifying glasses

Fiber Optics

Lenses

Laser pointers

Light sensors

Polarizing filters

Black plate

www.ingramcontent.com/pod-product-compliance
Lightning Source LLC
Chambersburg PA
CBHW080451220526
45465CB00006B/2231